Montessori & AI

Volume I

Kate Broughton

KATE BROUGHTON

Kate Broughton Ltd.

ISBN: 978-1-966665-08-3

Independently Published by Kate Broughton Ltd.

Book Cover Design by – Alex Broughton

For permissions, inquiries, or additional information, please contact:
Email: hello@katebroughton.com
Website: www.katebroughton.com

Published in the Unites States of America.

Contents

Chapter 1

Spark I

This book offers a series of musings, short, reflective pieces, designed to spark curiosity and inspire conversation. Through the lens of timeless Montessori principles, it explores the profound shifts brought by artificial intelligence. Each musing invites readers to pause, reflect, and engage deeply with ideas shaping the future of education and humanity. It is not a guidebook but a journey, encouraging readers to envision what is possible when technology and timeless wisdom converge.

Midjourney AI Prompts will appear under the images for each chapter, for example, "An HD wooden cube on a white surface, with a white background."

Chapter 2

Disclaimer

This book contains original ideas and content that represent solely the views and interpretations of the author. All material presented herein is the intellectual property of the author and is offered for informational, educational, and reflective purposes only.

The perspectives expressed in this work do not represent or reflect the views, positions, or policies of the American Montessori Society or the Innovation Fellowship Think Tank (2023–2025). Neither organization has endorsed, authorized, or assumed responsibility for any portion of this content. Any references to them appear solely to provide context regarding the author's professional experiences.

This book contains AI-generated images. While every effort has been made to ensure they are respectful, accurate, and contextually appropriate, AI systems may occasionally produce visuals that are imperfect, culturally imprecise, or not fully aligned with authentic Montessori materials. These images are included only to spark dialogue, reflection, and further inquiry. No offense or misrepresentation is intended.

The author disclaims all liability for any misunderstanding, misinterpretation, or perceived offense arising from the text or images. Readers should interpret all material as exploratory in nature and not as authoritative representation of any organization, culture, or pedagogical tradition.

"An HD photo of Montessori in the year 5,000 BC."

Chapter 3

Dedication

To my brilliant and creative family, Mark, Amelia, Olivia, & Alex, whose unwavering love and support have been my north star. To my cherished friends, mentors, and co-workers who have enriched and guided this journey with wisdom and encouragement, and to all the dreamers and innovators who strive to inspire and transform the world.

This book is for you.

May it light up your path as you have illuminated mine.

"The look of curiosity and wonder on the faces of a diverse range of children of different cultures while they explore nature in the classroom and interact with each other with joy and enthusiasm. Use hyper-realistic images with a DSLR close-up 50mm lens in the warm afternoon light coming through the window. The film should have realistic color rendering. We especially need imagination in science. It is not all mathematics, nor all logic, but it is somewhat beauty and poetry."

Chapter 4

Foreword

In February 2023 to march 2025, I worked with the American Montessori Society (AMS) as part of Cohort #1 & #2 of their Innovation Fellowship Think Tank. This think tank, concluding in March 2025 with the presentation of the Innovation Labs: The Future Montessori Classroom, at the AMS Montessori Event in Denver. This period and explore ideas in Montessori and Artificial intelligence (AI), unfettered by rigid rules and doctrine. In the spring of 2023, our small technology group began exploring how to articulate our society's position on technology and Artificial Intelligence (AI), mainly Open AI's ChatGPT 3.5, in April, which entered the world on November 30th, 2022. The conversation around technology became almost redundant as we grappled with the implications of Montessori and AI.

In April, I surrounded myself with newly minted AI newsletters popping up daily, technology industry brochures, and sparse academic publications to inform our explorations. The enormity of the task was palpable. Interestingly, I was in the unique position of having retired, which allowed me to explore these "Rabbit Holes," as Debbie Adams (IFTT) would describe them, without concern for my career. Our group, along with the AMS leadership team, soon began hosting Open Forum conversations on Zoom, with presenters Christian Pinedo (AiEdu), Jane Park (Google), Nita Farahany (Duke University), and Eric Rodriguez (Intel & IFTT Facilitator), and our precious Montessori community, to share their fears, professional viewpoints, hopes and dreams on these emerging technologies.

This book was born from my personal musings during my regular walks to the local recreation center. I would first hear the title, then words would randomly spill out of my mouth, all captured on a voice recorder app. These ten-to-fifteen-minute free-flow sessions would contain the essence of one idea, which I would then transcribe, albeit with imperfect accuracy, and email myself for safekeeping. As the months passed, I realized there might be value in sharing these thoughts in their raw, unacademic form, so they are for you, as one mind, heart, and soul to another.

These thoughts are not in any order, nor are they academically cited or referenced. They are simply a collection of reflections on AI and its impact on education and society. This book is for everyone, those new to Montessori and AI, and those deeply involved in Montessori or AI programming. With over forty years in education and most of that time in Montessori, I am committed to the Montessori pedagogy, which powerfully serves humanity during this critical time in our global history.

I had always dreamed of being an author, though I never imagined the process would feel like building while flying. This journey has been both exciting and energizing, though at times I have felt moments of paralysis, stuck beneath the weight of edits and revisions, unsure how to move forward. This was never intended to be a book. It began simply as my thoughts, yet they grew so quickly that I felt called to share them.

What you hold in your hands comes to you imperfect, and I offer it as it is. If you find any errors, as I once wrote in my *Mapping* book, I apologize. I have not taken this manuscript through a rigorous editorial process. Instead, I have leaned into new tools and companions along the way, using Alice AI for transcription, and Atticus as my editing platform. Each of these processes gave me the next step I needed, guiding me toward this moment of offering.

And so, with all its rough edges and honest beginnings, I share this with you, in love and gratitude. Thank you for receiving it.

Full disclosure:

I have leaned on Artificial Intelligence throughout this process, to record my spoken words, to transcribe them, and to help me shape spelling and punctuation into something legible. In May, the artist in me was captivated by the AI image platform Midjourney; it had me at "hello." Every image in these pages was born through that collaboration. I have done my best to pair the images with the sections, though at times I stretched my artist's license thin.

I offer these pages simply, trusting that you will use them in the way that serves you best, whether for reflection, conversation, or quiet companionship. Whatever wisdom you may find here is not mine alone; it has emerged in the shared space between us, as together we navigate this extraordinary moment in history.

"Step into a world where Montessori education and cutting-edge technology intertwine, creating a radiant vision of learning. Paint a picture where children's natural curiosity is nurtured in a Montessori environment enhanced by the power of technology. Explore the harmonious blend of hands-on exploration and digital tools, where children collaborate, create, and discover in an engaging and interactive way. Visualize the vibrant classroom where traditional Montessori materials seamlessly integrate with innovative digital resources, sparking a love for learning and igniting young minds with endless possibilities."

Chapter 5

How to Use this Book

This book was created from recordings made during the spring, summer, and winter of 2023 and 2024, in the early phases of artificial intelligence programs being noticed more by the public. Although AI has been around for many years, it entered the general domain more prominently when ChatGPT 3.5 made AI a focus of public and urgent discourse.

This book is a series of musings and questions I had during this time, related to but distinct from a think tank setting. These are my reflections and do not represent conversation within the think tank. The writing pieces are arranged in the order in which they were recorded.

It can be read in a straightforward book format, page by page, but you can also open it randomly and read whatever is there for you. The musings may be helpful if you are on a staff workday at a school with administration. You could open a page randomly, read it aloud, and discuss it as a theme. As a former department and school head, I would have been pleased to have such a light-handed resource on this weighty and sensitive subject. It could serve to discuss nuanced ideas about a topic, not in any strict order, but as a weekly process because everything is moving so fast.

So that you know, the Howard Gardner multiple intelligences breakdown is at the back of the book if you want to review that to give yourself a sense of where you are in the intelligences and what your strengths are. It is always good to have that in your awareness.

Please enjoy the book, take it from my heart to yours, and let us have conversations about artificial intelligence that genuinely makes a difference for humanity.

"How to use this Montessori Book."

Chapter 6

Midjourney Image AI

Before you step further into these pages, I want to invite you into a practice of observation. Look carefully at each image. Read the prompt written beneath it. Ask yourself: Did the image follow the prompt? If not, what went wrong? Jot down your ideas, your questions, your theories. Make notes in the margins. Research, reflect, revolt, it all works here. This is not passive consumption; this is your participation.

All of the images in this book were created using Midjourney AI. Since May 23, 2023, I have spent countless hours experimenting with this generative art program, learning its quirks, marveling at its brilliance, and laughing at its failures. I began on Discord, playing a kind of digital whack-a-mole as my prompts and images popped up and disappeared in a chaotic stream. Over time, the chaos became rhythm, the rhythm became mastery, and by the close of Midjourney's beta program in December 2023, I had generated more than ten thousand images. I had become what they called a "super user," swept along into their permanent program without fanfare but with deep gratitude for what I had learned.

Midjourney, developed by San Francisco–based Midjourney, Inc., is part of a larger revolution in artificial intelligence. Launched into open beta on July 12, 2022, it transforms natural language descriptions into vivid, often startling images—standing alongside OpenAI's DALL·E and Stability AI's Stable Diffusion as one of the most influential creative tools of our time.

Yet, as you begin to observe its creations, you will notice patterns that unsettle as much as they inspire. Hands often sprout too many fingers, bodies bend in strange directions, and faces fall back on narrow defaults unless pressed otherwise. Women are most often rendered as young, pretty, and white. Children, when they appear, are serious, white boys in uniform rows. Montessori materials balloon into absurd pyramids and endless stacks, stripped of their quiet precision. These distortions are not random glitches; they reveal something deeper, the biases hidden in the data on which the AI was trained.

Such biases are not unique to Midjourney. They reflect a broader truth about artificial intelligence: that these systems carry within them the imprints of our culture, our histories, and our exclusions. Joy Buolamwini, in her TED Talk *How I'm Fighting Bias in Algorithms*, calls us to see these not as trivial quirks but as matters of justice. To notice them is to join the growing work of algorithmic accountability.

So as you encounter the images in these pages, do not only marvel. Notice. Question. Ask what is missing, what is distorted, and why. Midjourney is not merely a tool for generating pictures, it is a mirror of our collective consciousness, a reflection of our biases, our imaginations, and our dreams. If we look carefully, it can help us ask not only what AI can create, but what kind of future we are willing to build.

"An HD, hyper realistic photograph of an elephant sitting next to an cyborg in a digital cloud in a sitting room."

Chapter 7
Dr. Maria Montessori

In a world experiencing social upheaval and rapid advancements in Artificial Intelligence, the legacy of Dr. Maria Montessori emerges not only as a historical testament but as a relevant beacon for today's education of children. Born in Chiaravalle, Italy, on August 31, 1870, Dr. Montessori broke through the societal limitations of her time as one of Italy's first female physicians. Yet, it is her revolutionary approach to early childhood education that marks her most significant contribution, much like the innovative disruptions we witness today in various fields due to AI.

Dr. Montessori's era, like ours, was one of profound changes and reevaluation of established norms. Her method challenged the rigid, teacher-centered educational practices of the early 20th century, introducing a child-centered approach that emphasized independence, creativity, and a love for learning, values that echo in today's shifting educational paradigms.

In 1907, the opening of the first Casa dei Bambini in Rome marked the beginning of what would become the Montessori method, a philosophy of education that views children as naturally eager for knowledge and capable of initiating learning in a supportive, thoughtfully prepared environment. This was a stark contrast to the prevalent education system that often-stifled creativity and imposed learning from the top down.

Montessori's classrooms were spaces where children, guided by their interests, engaged with specially designed materials, learning through hands-on experiences. This method fostered a sense of autonomy and self-motivation, which is vital in today's world where adaptability and critical thinking are paramount. Her work also extended to the formation of a deep partnership with her son, Mario Montessori, who was born in 1898. Together, they refined and spread her educational Method.

As we stand amidst technological transformations that question old paradigms and propose new ways of thinking and interacting, Dr. Montessori's educational philosophy offers a powerful reminder of the importance of human-centric approaches in education. It reassures us that in the face of AI and other disruptive technologies, the fundamentals of human creativity, community, and compassion continue to serve as our guiding stars.

For those new to her work, Dr. Montessori's represents more than just an educator; she embodies the role of a visionary who saw beyond her time. Her life's work invites us to reflect on our current educational practices and to innovate with mindfulness towards the development of the whole individual.

Dr. Montessori's story is a call to think critically about how we prepare our children for a future where they will navigate complexities possibly far beyond what we can currently foresee.

"A close-up shot of black, brown, red, white, and yellow, middle and high school students, all genders, in a clean, clear, randomly spaced, light environment, working with Montessori materials. One child is using a laptop and creating a holographic image. Use a DSLR and a 50mm lens, take in the soft light of the morning with Kodak Portra 400 film. The human hand, so delicate and so complicated, not only allows the mind to reveal itself but it enables the whole being to enter into special relationships with its environment."

Chapter 8
Short AI History

Long ago, when the world was simpler, a group of thinkers dreamed of creating machines that could think like humans. They imagined mechanical beings with human-like intelligence capable of reasoning and learning. Little did they know that their musings would one day give birth to a field that would change the world forever: Artificial Intelligence.

As the 20th century dawned, the stage was set for AI to take its first steps. A brilliant mathematician, Alan Turing, asked, "Can machines think?" His ideas laid the foundation for the Turing Test, a way to measure machine intelligence. At the same time, the first programmable computers, like the ENIAC, were born, giving AI the tools it needed to grow.

In 1956, a group of scientists, including John McCarthy and Marvin Minsky, gathered at Dartmouth College for a historic meeting. It was here that AI officially came to life as a field of study. The excitement was palpable as they dreamed of creating machines that could reason, learn, and communicate like humans.

Over the next few decades, AI began to show its potential. Programs like the Logic Theorist and ELIZA demonstrated that machines could perform tasks that seemed uniquely human, like solving problems and engaging in conversation. Hopes were high, and money poured into AI research, fueling a period of rapid progress.

But the road to AI was not always smooth. In the late 1970s and early 1980s, researchers hit roadblocks. They realized that creating accurate human-like intelligence was much more complex than imagined. Funding dried up, and many began to doubt whether AI was possible. It seemed like AI might be nothing more than a pipe dream.

Yet, just as all hope seemed lost, AI experienced a resurrection. In the late 1980s, new approaches like machine learning and artificial neural networks emerged, giving AI a fresh lease on life. Computers could now learn from data and experience rather than just following predetermined rules. It was a game-changer.

As the 21st century dawned, AI exploded into the mainstream. Powered by massive amounts of data, deep learning algorithms, and exponential increases in computing power, AI began to achieve feats once thought impossible. It could recognize faces, understand speech, and even drive cars. AI has become a part of our daily lives from smartphones to hospitals.

Today, AI continues evolving at a breakneck pace, transforming industries, and shaping our world in ways we never imagined. The story of AI is one of dreams, challenges, and incredible achievements. It is a testament to human ingenuity and the power of perseverance.

"Piece of paper with rough sketch of a Turing machine."

Chapter 9

Musings

As I have been editing these musings, I have had to override the feeling of needing to justify different comments, edit sections, put them in order, or ensure they follow a logical sequence. I have felt the need to make sure I do not offend anybody, ensure it does not sound foolish, and maintain an appearance of intelligence in my thoughts. I have strived to cover all bases and areas of interest, confronting the ingrained school training on how a book and its chapters should be structured.

The idea of fixing a mere thought, editing it to sound more intelligent, and adding explanations is deeply rooted in my educational background. Within these musings, I have realized the immense freedom, rest, and peace it has brought to my thinking. I did not feel compelled to do anything other than allow the thought to emerge and be acknowledged without justification or alteration, honoring the thought so that it felt seen and heard.

Ironically, this is akin to our goal of ensuring children are seen and authentically heard, which suggests that a thought can be an entity deserving to be acknowledged as it is without adjustments or justifications. I wonder, for you, the reader, when thoughts flow through your mind, not fears, worries, or concerns, but thoughts about life and questions, do you experience a childlike sense of wonder?

This wonder about how the world works, how people behave, and the origin of our knowledge is not childish but a state of pure curiosity. I ponder how school systems have evolved, how children adapt in the age of AI, and the nature of our intuitive thought processes.

Could a softer, more wondering approach facilitate better interactions, especially on complex topics like AI, politics, and religion? In a time when emotions can easily lead to fear and judgment, I propose that making space for gentle musings could foster a unique way of interacting, where we can be authentic and heard without the need for editing or restructuring our thoughts.

This is a pivotal moment in history where certainty is fleeting, and what we know may only be true momentarily before the world shifts and new understandings emerge. In this context, wondering, rather than claiming to know, may be our best approach.

"Step into a world where Montessori education and cutting-edge technology intertwine, creating a radiant vision of learning. Paint a picture where children's natural curiosity is nurtured in a Montessori environment enhanced by the power of technology. Explore the harmonious blend of hands-on exploration and digital tools, where children collaborate, create, and discover in an engaging and interactive way. Visualize the vibrant classroom where traditional Montessori materials seamlessly integrate with innovative digital resources, sparking a love for learning and igniting young minds with endless possibilities."

Chapter 10

Montessori's Quantum Jump

The phrase "Follow the Child," central to the Montessori method, resonates deeply with me, urging me to rethink our roles as educators in an age where quantum physics plays an active part. The Montessori method positions the child as the focus of our educational practices, where we understand that just the act of observing the child can significantly shape their learning path.

Dr. Montessori's creation of, and work at, La Casa dei Bambini was revolutionary. She crafted an environment with child-scale furniture and unadorned walls, aiming to observe children in their natural state. The method after her name, seemingly straightforward, was a complex exploration of child development. In today's world, we face challenges in authentically replicating her approach, considering modern changes in teacher training, classroom dynamics, and true educational access.

Our task today is to apply Montessori's principles across global education, challenging conventional methods and structures. This may involve returning to the Montessori basics of minimalism and simplicity, even in diverse educational settings. How do we maintain a clear, clean space conducive to learning, whether indoors or in nature, while being true to the physical location and digital age in which we live?

A crucial factor is the training, and the intent of the adults involved. How do we align our classrooms with Montessori's vision to observe and understand children authentically, without rigid interpretations that exclude diverse communities? We must evaluate if our Method fosters self-expression and normalization for all children.

As we adapt Montessori education for the digital era, how do we keep it child-centered and respectful and integrate technology? This evolution is not just about preserving a legacy but about adapting it to maintain its profound impact on future generations. The challenge is to reflect on its origins, understand its core tenets, and evolve to meet today's realities, thus continuing Montessori's significant influence on education.

"An HD image of a classic acorn nestled in leaves on a forest floor in the afternoon."

Chapter 11
Off Kilter

It is interesting how easily we, personally and as a world, have been thrown out of kilter with the arrival of artificial intelligence. Observing different business environments, schools, and everything around me, I have realized that it is almost like we have been knocked off balance. Whatever we had been doing relatively successfully until COVID-19, the systems that worked and the processes we were accustomed to using all seemed fine. We thought everything was humming along nicely. It may not have been perfect, but it was manageable, and we were doing our best. The idea of "life as usual" seemed straightforward. We had our fingers on life's buttons, dials, and levers and felt a degree of control.

Since the arrival of artificial intelligence in public discourse, it is fascinating how we may now feel knocked off center. It is as if there is no way we could be doing everything right because if we do not consider artificial intelligence we would be missing out on crucial advancements. For businesses, the thought process shifted almost immediately. By June, after artificial intelligence became a topic of conversation in April, we felt the need to have something in our school handbooks to acknowledge our awareness of it. Knowing that since 2019 and even earlier, businesses have had artificial intelligence on their radar and have already included it in long and short-term planning and had initiated pilot schemes, it is intriguing how it remained under the radar for so long but has now become a challenge for everyone to address.

Being knocked off kilter is an emotional state of being caught off-guard and trying to figure out what to do next. Running any organization, leading a school, or managing a classroom involves many tasks. There was hardly any space for this new consideration to be dealt with. In addition to all the regular training in schools, such as curriculum planning and social-emotional development for children, we are now faced with integrating artificial intelligence. This new requirement must now rise to the top of the priority list.

We may now have this feeling that something needs to be done about artificial intelligence, but I am wondering what we could bring to the discussion that does not add to the sense of urgency. Instead, we could bring a deliberate sense of peace and calm, reassuring that children and schools are not going anywhere. We already have the fundamentals in place. Maybe the conversation could focus more on what we should do rather than how quickly we can deploy artificial intelligence or ensure everyone can access it. By remaining focused on the child's well-being and observing their needs, we are unlikely to go astray. A peaceful, deliberate, methodical, and science-based approach should guide what we introduce into the classroom for our precious children, ensuring their best interests are always at the forefront.

"Montessori teachers in a workshop setting: Continuous training for educators partnerships with tech developers community involvement."

Chapter 12

Talking to Machines

As I reflect on the evolving realms of education and technology, a pressing question emerges: Are children eager to learn how to interact with machines? This is especially relevant for the Montessori community, where we have traditionally balanced technological advancements with the Montessori ethos of a hands-on, sensorial environment.

Our cautious approach to technology, aimed at preserving the child's learning space, reflects our commitment to their development. Yet, considering technology's pervasive role today, is it time to reevaluate our stance? Children are growing up in a digital era, where technological fluency is as crucial as basic literacy.

If the Montessori community takes the lead in shaping technology use, particularly AI, can we redefine its role from potential disruptor to an enhancer of the Montessori method. We can safely manage the challenge of integrating technology in a way that complements, not dictates, the child's learning journey.

Envision AI not as a mere tutor but as a supportive guide in the learning process. This AI would respect the Montessori principle of following the child, aiding in identifying areas of interest or confusion, and guiding exploration at the child's pace.

Our voice in technology may be crucial, especially as we already witness children's comfort and proficiency with technology. Ignoring technology's role in education would be out of step with the realities our children face. However, if we integrate it thoughtfully, prioritizing developmental needs, we can revolutionize education.

Our community, including tech-savvy educators and forward-thinking administrators, recognizes the importance of engaging in technology discussions. It is vital that we participate, advocating for educational technology that respects and fosters physical and cognitive development.

Including children's perspectives in these dialogues ensures that technology aligns with their learning needs, rather than adult assumptions about educational tech.

Faced with a critical juncture, the Montessori community must decide whether to proactively adapt to technological changes or remain reactive, potentially missing the opportunity to influence a key aspect of modern education. This moment calls for thoughtful innovation and reaffirmation of our core educational principles.

"A photograph of a light bright Montessori middle school classroom with 25 active and hardworking students, desks are arranged in small groups, wheelchair access is in the center of the room. There is a wide range of materials and manipulatives available to all."

Chapter 13

Should we be Frightened of Our Children?

The question is, should we be frightened of our children? An exciting thing right now is to know that AI has been influencing our world for years. In the classroom, we may be seeing evidence of what our children are doing with AI. Are they using it to cheat on tests, essays, homework? The bright kids, the ones we could be frightened of, could be using AI for their own purposes, employing it to circumvent rules and short-circuit processes, to ease their own academic stresses so that they can focus on things that are more important to them. Incentivized individuals may already be making money from their AI use, crafting cool AI products to sell. There has never been a shortage of people in history who quietly seize opportunities in times of social change and upheaval.

What we have right now with AI is a system that may be used for the worst incarnations of human nature without regulation, matching up perfectly with a student body who have disassociated from the school system, because of its incessant pressure for performance regardless of ability, engagement, enjoyment, or meaningful progress. It is performing at all costs, and we are exhausted. We are seeing this in the world of work, where adults are in an arms race for information and paperwork just to stay ahead. We have now been run over by the freight train of AI and we are playing catch-up. We cannot foresee the future because we are still trying to prop up an outdated education to work pipeline.

This system is falling apart. Teachers are already at a disadvantage, and the children are trying to survive. Add to this mix, children who are feeling unheard, unseen, angry, frustrated, or depressed, and then give them AI, a tool that does not judge them or roll their eyes with impatience, and you have a terrifying situation.

Should we use outliers, malcontent children, and wounded souls to decide our future with technology and what our schools should do? What if we considered a different approach? In Montessori education, it is well established that we follow the child. Children have freedom of movement, thought, and conversation. When you are zero to six years old, you learn to find your way with others because you value community, being heard, and being seen for who you really are, and these values affect your personal direction for years to come.

This all sounds good, and you might say, look at Larry Page, Sergey Brin of Google, and Jeff Bezos of Amazon – all had Montessori in their educational portfolios. Yet, if that approach worked, wouldn't we be spared the nightmare of intrusive technology like Google's? It makes me wonder if the combination of a detached and desensitized child population and this AI weapon of mass distraction is why we are seeing an apocalyptic view of the future rather than a beautiful, integrated, semi-organic future where AI is used to support humanity, not destroy it.

The question remains: Should we be frightened of our children?

"A close-up of a happy and inquisitive young girl sitting in a Montessori classroom, the perfect blend of AI, technology, and the natural world."

Chapter 14

AI & Learning

In his book "Life 3.0" by Max Tegmark, he discusses different learning styles and the unique capabilities of AI. He notes that AI can have a supreme understanding and awareness of one specific item or task due to its focused nature, but it lacks general abilities and awareness. This led me to think about Montessori education and what captures children's attention in the classroom. Specifically, I pondered the valuable aspects of classroom experiences that children genuinely love. In my own elementary classroom, I observed that the children cherished the respect and autonomy they were given. They relished the quiet time in the morning reading, followed by the group morning meeting. They appreciated my trust in them to create and follow their work plans, giving them pride in running the classroom.

Comparing this with public, non-Montessori schools, a stark difference emerges. In Montessori schools, children enjoy respect, autonomy, and freedom. In contrast, public school systems may offer pockets of these elements, but the rigid structure often limits deep engagement and autonomy. The bell ringing signals a shift to the next class, dictating the day's activities rather than allowing children to choose their studies.

Reflecting on AI and its role in education, I ponder how we can keep children engaged and joyful in any classroom setting. Our current technological environment, with its addictive nature and platforms like Facebook, Instagram, TikTok, and video games, can detract from educational engagement. If we remove external stressors like high stakes testing and predictive learning and focus instead on the child, the teacher, and the child's natural interests, we can regain their trust and interest in being part of a learning community.

Turning this system around may be like maneuvering an educational supertanker, but it is necessary. We must question where the truth lies in education and who holds it. Integrity, honesty, and transparency are crucial in our school systems, offering an alternative to the AI and online world. School must be a destination, a place of sanctuary for children, rather than something to be avoided.

We have a history of joking about school, but what if we talked about reparations for the harm the educational system has caused to children? Imagine a wave of claims, not because teachers or schools are inherently bad, but because the system became a trap, focused more on filling factory jobs than nurturing children. It was a system that devolved into prediction, routine and numbness for staff and students alike. Now, we can rewrite the narrative. This is not about change for the sake of change but because we clearly see the need. If our schools are not sanctuaries for children, then we are failing them, and that is not what we want to be doing in an age of unpredictability where if we do not influence their learning for good, then we have already washed our hands of their future.

"A culturally diverse Montessori child and a rabbit hole in nature."

Chapter 15
They Were all Children Once

As I reflect on the state of AI and its immense influence on our lives, I cannot help but think about its dual nature, highlighting humanity's best and worst aspects. It is May 7th, 2023, and AI shows us our motivations, hopes, dreams, and relationship with power and intelligence. AI is a product of children over eighty years of brilliant, inquisitive, and creative minds. Adults now, once children, who grew up in various educational systems, are products of their time. These children may have been inspired and supported by their own school experiences, or not; whatever happened their experiences were the fertile soil for AI's development. From those early seeds, subsequent generations have built upon the wisdom, creativity, and investment of the past. With AI, we see the culmination of decades of creativity, standing on the shoulders of giants. The AI's we see today were built by these children.

To understand and engage with AI, we need to start with children. We should be alongside them, listening to and observing how they interact with AI. Imagine if every classroom, starting with Montessori but extending to all, incorporated AI as a scientific experiment, where children were treated as scientists, their opinions listened to and integrated into future iterations of these programs.

What if we presented children with an open-hearted, powerful request for help? We adults do not have all the answers, and it is crucial to know what the children, the future bearers of our world, desire. The notion that we are protecting children by excluding them from these discussions may be misled. If you ask an elementary child to plan the rules for their classroom, they can do so effortlessly, producing rules about respect for others, the environment, materials, time, and nature. This is easy work for them.

Involving children at essential levels in discussions about AI could be transformative. Imagine if young Larry Page or Sergey Brin of Google were asked for solutions at nine or ten years old; would they not have produced something remarkable? Can we approach AI research with humility, kindness, and honesty, and earnestly ask our children for help? Let us stand beside the new generations of giants, and see them for who they really are, whole and complete right now, dribbles, mess, and all.

"Diverse child exploring pond scum with a bucket and net, in an HD hyper realistic close-up image. The sun is shining on the water."

Chapter 16

Intersections

What an intriguing intersection, the warm nurturing of the Montessori classroom meeting the cold calculation of artificial intelligence. My mind spins with possibilities even as my heart urges caution.

I picture our little ones fully immersed in their lessons, tiny hands practically glowing with the magic of discovery. They beam each time they master the next step, intrinsically celebrating growth not for some external award but for the pure joy of rising to meet their budding potential.

Then I envision the emotionless algorithms of AI, seeking not to uplift the human spirit but to optimize statistical efficiencies. And yet, might such technologies have their place in liberating human potential? Carefully implemented, could they handle administrative minutiae and rote drilling, freeing up precious time for teachers to encourage deeper question-seeking? Rather than replacing human guidance, might AI thoughtfully assist it?

I watch as our children learn to self-direct their development through prepared Montessori environments. They thrive in this protected greenhouse that nurtures independence equally with interconnection. Step-by-step they construct not just academic knowledge, but the very skills needed to flourish in a tech-driven world - inquisitiveness, discernment, communicative collaboration.

Truly, our Montessori students organically build the executive functioning and adaptive flexibility that AI systems themselves need. Our little ones create order while following innate passions, fully immersed in the joyful flow of learning. They code their own operating systems through hands-on materials and the freedom to fail without judgment.

I feel hopeful that this natural dance of curiosity and structure plants seeds of self-construction that should serve our children well as AI proliferates globally. But we must guide these technologies cautiously, shaping their integration in human spaces centered first on compassion over efficiency or profits.

If she were alive today, I believe Dr. Montessori would remind us that the most vital preparation for any future is to nurture humanity within each child now. There is light ahead if we walk this path together, discerning each step with patience and with our full humanity present.

"A close-up shot of black, brown, red, white, and yellow children working with Montessori materials and nature, using a DSLR and a 50mm lens, taken in the soft light of the morning with Kodak Portra 400 film. The human hand, so delicate and so complicated, not only allows the mind to reveal itself, but it enables the whole being to enter into special relationships with its environment. In the background, have children chatting with each other and others reading in a reading corner."

Chapter 17
Assumptions

I find myself contemplating the complex world of AI and the assumptions that I make about life. It is May 7th, 2023, and AI seems to accentuate both the best and worst of human nature. I realize that my assumptions shape my experience of life. Assuming the world is a kind, safe place brings me peace and comfort, letting me believe that my family is safe, our needs are met, and our home is secure. But what if I assumed the opposite? If I viewed the world as unsafe, full of danger, and devoid of support, my life would be filled with fear and distrust.

Our assumptions indeed shape our reality, but what about when it comes to AI? People believe we are far from an AI that could dominate humanity, seeing it merely as a program without inherent good or evil. Yet, there are fears too, assumptions that AI will take over jobs, rendering human skills unnecessary. While others believe that AI will free up our time for better pursuits, while others fear income loss and a future where traditional jobs are scarce.

These assumptions extend to the school system. In the past, there was an emphasis on testing and compliance, with a belief that students could not manage themselves without strict rules and monitoring. This approach fails to trust children's innate capabilities and motivations. Now, children with access to AI technologies like ChatGPT are learning beyond the traditional school system. They are exploring new opportunities and opening previously unavailable realms of creativity, but what happens when this enthusiasm meets a school system that might not welcome or understand it?

Our fears and assumptions as teachers, parents, and students create a gap, an unoccupied zone, between potential and reality. In the Second World War, soldiers from opposing sides met briefly in a no-man's land to connect as humans in a holiday period, setting aside their differences. What if we created a similar space in education? A space where assumptions could be suspended, and we connect as equals to understanding AI's potential and the realities it could bring. This space of creation and conversation would exemplify the best nature of a think tank, with the parts benefiting the whole.

In this space, we could combine the wisdom of teachers and the fresh perspectives of children stepping into this world as their future. They could have a significant role in shaping how AI is used in education, balancing its use, and helping regulate its integration.

It is time to let go of our assumptions and see what unfolds. Let us embrace this opportunity to listen, learn, and co-create a future with AI that respects both the legacy of teaching and the innovative spirit of our children.

"A close-up shot of an indigenous child's hands interacting with natural materials, using a DSLR and a 50mm lens, taken in the soft light of the morning with Kodak Portra 400 film. The human hand, so delicate and so complicated, not only allows the mind to reveal itself but it enables the whole being to enter into special relationships with its environment. In the background, have children chatting with each other."

Chapter 18
Montessori Normalization

I am contemplating the concept of normalization within the Montessori environment. This term symbolizes the harmonious, peaceful state of a classroom where children of all ages, from infants to secondary school students, are deeply engaged in their self-chosen work. This state of normalization, characterized by focus, freedom, and ontological being, results in a calm, orderly environment, a kind of magic that we Montessori educators aspire to. It signifies that the prepared environment, materials, and lessons in grace, courtesy, and self-care have nurtured content, engaged, and happy children.

However, as we venture into the Human Brain Interface (HBI) and Artificial General Intelligence (AGI), I ponder what human magic looks like in this new context. Could it resemble students engrossed in laptops, immersed in creating stories or projects through platforms like ChatGPT? This image of children deeply engaged with AI may appear to be a form of normalization. AI's ability to generate brilliant, evocative content can be incredibly captivating and seductive.

Yet, this raises a critical question: Is this truly normalization, or is it an illusion? When children opt for AI over traditional Montessori materials, are we witnessing a genuine engagement or substituting one form of learning with another potentially less enriching? How does this shift impact the value of hands-on learning experiences like geography maps or biology experiments, which could now be replaced by computer-generated simulations?

In the broader context of education, this shift raises questions about personalization and bespoke learning experiences. Traditional education systems often focus on delivering a standard parcel of information, regardless of the child's personal interests or desires. Montessori education, however, offers a personalized approach, allowing children to follow their interests and create projects that reflect their individuality.

The introduction of AI in classrooms might initially manifest as an addiction or a profound seduction for children who have not been historically satisfied by traditional methods. Immediate gratification and the ability to express themselves creatively through AI can be overwhelmingly appealing. This might lead to children preferring AI over traditional methods, foregoing recess, or snack times, engrossed in the captivating world of AI-generated stories and projects.

However, this raises a crucial question about preparedness and success. Have we adequately prepared our children for this transformation? The hunger for self-expression and validation might lead to an overreliance on AI, which, while empowering, could also overshadow the importance of independent thought and creativity.

Ironically, the creators of AI have provided a platform for children to be seen and heard worldwide. The challenge for Montessori educators is to find a way to integrate AI while prioritizing Montessori principles in a balanced and educationally enriching manner.

"Soft authentic splash mark for a sketchbook."

Chapter 19

Slow AI

The concept of superintelligence has led me to ponder the need for programs that effectively manage information. Our educational environments in schools and workplaces are incredibly information dense. This density, evident in continuous assessments and evaluations, has become accepted. We are inundated with reports, data management, and documentation, reflecting a world characterized by endless information processing. The workplace has often been viewed as a joyless, sensorily deadening environment, filled with red tape and bureaucracy, a perception we may for the most part have just accepted.

Imagine a bespoke Montessori AI platform based on Montessori legacy information, all of Dr. Montessori's writings, information held in universities' research institutions, and all the lessons as initially designed. We may instantly see conflicts or biases once this information is uploaded. In fact, it is hard to imagine what we would see. I am compelled to think about how global Montessori organizations can collaborate closely with programmers, educators, and historians to invent a Montessori AI platform that powerfully forwards Dr. Montessori's intent. I would be particularly excited by an AI program that adds the grammar symbols above the words as the student creates their story. AI sentence analysis would be out of this world.

Today, Montessori students are immersed in experiential environments where everything is purposeful and meaningful. They continuously add to their physical, mental, and emotional data banks through experiences rich in taste and texture. In contrast, traditional education often inundates children with meaningless work, coloring sheets, and excessive homework, mirroring the adult workplace's inefficiencies. As Montessori educators, how can we leapfrog this hurdle and redefine the 21st-century workplace, especially with AI in the picture?

Our goal could be to integrate AI not as a dominant force but as an invisible tool, aligning with the Montessori community's ethos. We could envision a platform so seamlessly integrated into the classroom that it is unnoticeable, akin to how we use Word documents or Excel – always there, available, but not obtrusive or compulsory.

Children learn through physical and kinesthetic experiences in a Montessori classroom. During morning meetings, students use an old, well-loved dictionary to research words, oversized, leatherbound, with gold embossing, smelling of printing ink and musk, the pages stained with the oil of little scrolling fingers. This rich sensory experience forms deep emotional connection to the information gleaned; something that may be relinquished with the convenience of the digital assistant.

What about 'Slow AI' mirroring the Montessori approach of slowing down and allowing children to immerse deeply in their activities? Slow AI would be about supporting the educational process without compromising the depth of learning. It is about advocating for a balanced integration of AI that respects the child's need for authentic engagement.

"A photograph of a serene landscape using your DSLR and a close-up 50mm lens during a peaceful morning. Use Kodak Portra 400 film to capture the true essence of the environment. This image should embody peace, nature, the beauty of Montessori principles, and childhood innocence."

Chapter 20

Follow the Child

I have been pondering the integration of AI in Montessori environment. As I delve into this, I think about the principle of 'Follow the child' in a world increasingly dominated by technology and AI. Traditionally, in Montessori settings, normalization is achieved through a series of powerful processes that lead to an engaged, peaceful, and content child. This state is not forced but emerges naturally in a prepared environment where everything a child needs is available, from sensory materials to math and language resources.

However, as we introduce AI into these environments, we face new challenges. The Montessori approach allows for a certain degree of freedom, but there is also a need for guidance and direction. For instance, a teacher might use subtle cues to help children focus or complete tasks. This balancing act becomes even more complex in an AI-enhanced world.

The traditional school system often lacks this balance in the broader educational landscape. Montessori children often struggle in non-Montessori public schools. They arrive in an environment with limited physical movement and agency, and they are suddenly expected to conform to a rigid structure alien to what they experienced in a Montessori setting. This transition can be jarring as they encounter distinct social norms, values, and behaviors.

With technology, particularly social media platforms like Facebook, Instagram, and TikTok, children find a different kind of reinforcement that feeds on their need for social and emotional validation. As educators, we observe children not choosing activities that we know they would typically be drawn to. Instead, they are increasingly drawn to the instant gratification provided by technology.

This shift poses a dilemma for Montessori educators: How do we maintain the essence of 'follow the child' when the child is increasingly immersed in the digital world? This challenge is further complicated by the realization that schools are not prioritizing emotional connections. A reset is needed through initiatives like no-technology weeks, where the focus shifts back to connection, discovery, and personal interests.

In the public school system, and sometimes even in Montessori settings, we must ensure that children genuinely trust that we support and follow them. Having experienced the transformation of children moving from the public-school setting to the Montessori classroom, I have seen their amazement when they realize that they really have the freedom to move, to rest if needed, and to explore at their own pace. They learn the value of honesty and self-reliance, realizing that shortcuts like peeking at answer cards are unfulfilling.

In this landscape where AI is becoming increasingly prevalent, our approach may be not just about following the child, but about walking beside them into a future where technology and human values coexist in harmony.

"An HD, simple, highly muted graphic design of six plain wood cubes, to the left of the picture, balanced in a wild way, in front of a pale blue, white, grey, and cream background."

Chapter 21
Polarization

As I ponder the current polarization in American society, I am struck by the profound impact of social media platforms like Facebook, Instagram, TikTok, Twitter, and others. These platforms have amplified our collective urgency to assert our beliefs, draw a line in the sand, and passionately and persuasively defend our viewpoints. It is as though we are in a state of constant reaction, akin to stroppy teenagers or eight-year-olds at their developmental extremes, demanding, bullying, rude, and disconnected from a sense of community or shared reality.

This polarization manifests in two distinct ways. On one side, we have individuals who assert their "rightness," who demand to be heard and seen, and who are willing to circumvent laws and social norms to make their point. Conversely, there is a more passive approach where people disengage, believing this intense societal conflict will somehow resolve itself without their involvement. Both approaches, whether aggressive assertion or passive withdrawal, stem from a society that seems to have lost its connection to collaborative dialogue and mutual understanding.

When seeking the roots of this polarization, we can look at family dynamics, parenting styles, and educational experiences. Did we learn in our homes or schools that our voices did not matter unless we shouted or fought for attention? Did we grow up unheard, leading to a present-day urgency to reclaim our voice and power?

This situation raises the question of whether our educational and family systems have inadvertently contributed to this societal division. Have we, as a society, failed to provide our children with the sense of being seen, heard, and valued? If so, the path to healing and de-escalation lies in rethinking our approach to education and parenting.

Could the solution be as simple as listening more to our children, giving them a voice and autonomy, and fostering an environment where everyone feels seen, heard, safe, and loved? A population that grows up feeling valued and understood is less likely to feel the need to aggressively stake claims or disengage from society. By nurturing these values from childhood, we might just find the key to healing the deep divisions that currently characterize our society.

"Deleting notes x Regia seedpod."

Chapter 22

Standing on the Shoulders of Giants

In our world, there is a common misconception that every new idea or creation is born in isolation, unique, and unprecedented. We often forget that we are standing on the shoulders of giants, building upon the work and wisdom of those who came before us. This notion resonates deeply with me as I reflect on my own lineage and its influence on my perspective.

My grandfather, John Gow, was an integral part of the early days of matte projection, contributing to groundbreaking work as an art director at Pinewood Studios and having memorable encounters with icons like Metro Goldwyn Mayer. My father was a meteorological officer, engrossed in numbers, creating important rainfall tables still in use today. These family connections have shaped my worldview and instilled a profound respect for the continuous flow of human knowledge and achievement.

In education, both in the UK and the US, I have seen the ebbs and flows, the booms and busts of new curricula and educational systems. Plans implemented saw a significant improvement in both access and content of a good education. While programs could be an incredible and profound addition, others were detrimental to the students and the teacher regarding motivation and creativity.

Throughout the years, educational visionaries like Dr. Montessori and her peers persisted despite the oscillations in credibility and interest. They saw the potential for their ideas to significantly benefit humanity and pursued them with relentless dedication.

This concept of legacy and building upon past work is especially poignant when thinking about our children, not just those in Montessori environments but every child. Their growth, drive, and eventual contributions to society are the fruits of an extensive line of influence and inspiration. Professionals across various fields have experienced the delayed gratification of not seeing the immediate outcomes of their work. It reminds us that our advancements and achievements are not sudden or standalone but the result of cumulative effort and collective wisdom.

As we engage in conversations about AI, especially in the context of Montessori education, we can take comfort in grounding ourselves in this historical perspective. We can recognize that what we see today, the applications and implications of AI, are not spontaneous developments but the blossoming of ideas that have been nurtured over time. It is also important to acknowledge that while AI has become more pervasive, its benefits are not yet universally accessible. The democratization of technology remains an ongoing challenge.

Moving forward, we must ask deeper questions and show genuine appreciation for the innovators and programmers whose cosmic purpose has brought these advancements to fruition. We can advocate for the aspects of AI that genuinely support and care for humanity and our existence on Earth. We are not just recipients but custodians of a legacy, continually advancing the work of those who have paved the way for us.

"A hyper-detailed close-up of Leonardo Davinci's Sketchbooks showing his brilliance, creativity, and true essence."

Chapter 23

Struggling with the Future

As I delve into the complexities of AI and its impact on our society, I am struck by the overwhelming sense of uncertainty and upheaval. Everything we have known and held onto, from work norms and salary structures to relationships and safety, seems to be in flux. AI's burgeoning power is both alarming and exciting. It reminds me of my childhood studies of the Industrial Revolution, where the promise of more leisure time was offset by increased accessibility to material goods, altering lifestyles and freedoms.

In this new world, the creation and consumption process has become incredibly rapid and accessible. You can design a logo, create a website, and have promotional materials ready in minutes. While the new shiny thing, this ease of access raises questions about the value and uniqueness of individual creativity. When everyone can easily generate content and products, what happens to the concept of originality and the artisan's connection to their handiwork?

This shift in how we create and consume has profound implications for employment and lifestyles. Growing up in this digital landscape, the emerging generation navigates this world effortlessly. But as the baby boomers retire, and traditional job structures change, what will the future of work look like? Millennials have roundly rejected the conventional nine to five workdays, opting for more flexible lifestyles. The COVID-19 pandemic experimented with the concept of universal income, hinting at a potential shift in how we perceive and value work and money.

As we contemplate the future, we might consider how our human value has historically been tied to our productivity. The new questions might focus on creativity, passion, and dreams, echoing Montessori's emphasis on cosmic purpose. Yet, in traditional education systems, where students are often taught to conform and complete tasks without delving into their passions, how do we foster a society that values individual dreams and aspirations?

As everything is thrown into the air, how these pieces fall back down will depend on our collective vision for the future. Do we envision a dystopian world dominated by virtual reality and apocalyptic scenarios, or do we strive for a future filled with life, great food, and fulfillment? What would young people, especially millennials, envision as their ideal future? Understanding their aspirations could be key to shaping a world where AI is a tool for enhancing human creativity and potential, rather than a force that diminishes our sense of purpose and individuality.

In grappling with AI, we cannot lose sight of our own power and agency. When we dream big and put energy behind those visions, we can shape a future that aligns with our deepest values and aspirations. The challenge lies in thoughtfully and methodically harnessing AI's potential while staying true to our human spirit and creativity.

"Follow the child in AI, close up, diverse group."

Chapter 24
Having it all Together

In today's world, particularly in Western society, there is an overwhelming expectation of having it all together. This notion extends to every aspect of life, from finances and relationships to work and the future. Being 'together' is often equated with success, but this can lead to the suppression of emotions and a fear of failure. It is a systemic issue that can be seen in our public-school systems, where children who express their emotions or behave differently are often labeled and medicated.

In Montessori environments, we aim to foster children's normalization, autonomy, and self-actualization. They are encouraged to explore and learn at their own pace, in a space that respects their individuality. However, even within Montessori settings we are not perfect. There can be moments where teachers unwittingly exert control rather than allow the children to express themselves freely. This could be through subtle cues or expectations, children to conform to an ideal of "having it all together."

The advent of AI has introduced a new dynamic to this scenario. AI programmers and scientists have created systems capable of wondrous things, yet they are often unsure of how these systems work or the extent of their capabilities. This mirrors the broader feeling in society coming out of COVID-19, where people still feel they do not have it all together, compounded by fears of job loss to automation or being undervalued in the new digital landscape.

This situation calls for a more open conversation about our collective uncertainty and the need to reassess our approach to education and work. The traditional factory education model has been slowly transitioning into the digital age but has not kept pace with the changes around us. We might have missed following the child's natural curiosity and development in trying to control the educational process.

In Montessori education, following the child is not about letting them do whatever they want; it is about observing them and providing the materials and opportunities for their next natural developmental step. However, we must acknowledge that sometimes we do not know the next step.

We have seen children retreat into gaming as a reaction to educational systems where they feel unseen and unheard. It is seductive to enter a world where you are in control, and your actions have immediate, gratifying outcomes. So, can we partner with children to explore AI, acknowledging that we are learning together? This partnership means admitting that we do not have all the answers and embracing the fact/reality that none of us truly has it together. As we navigate this unfamiliar territory, it is crucial to maintain the core aspects of Montessori education that work beautifully; the hands-on, child-led approach. It is about understanding where the miracle of Montessori meets the challenges of the modern world and walking that path together with our students.

"A photograph of the factory model of school, the classroom has children behaving like robots."

Chapter 25
Extraordinary Skills

In today's world, where AI is rapidly reshaping our landscape, the call for extraordinary skills is louder than ever. While most of us possess basic, everyday skills, the AI age demands abilities that stand out in neon brightness. Beyond the essential technical competencies like computer programming, the AI era values and requires a creative mind, asymmetric thinking, and the art of Socratic questioning.

Individuals who excel in the AI age may not be just those with traditional technical skills but also those with the ability to think creatively and question deeply. These skills enable one to engage meaningfully with complex ideas, follow intricate thought threads, and take decisive action based on these explorations. In essence, the ability to think differently and innovatively is becoming as crucial as technical knowledge.

The looming concern in this evolving landscape is the potential loss of jobs traditionally held by what we might call the 'ordinary worker', roles that can be easily automated or taught to a computer. This shift poses a critical question: how do we prepare our children and ourselves for a future where extraordinary skills are not just an advantage but a necessity? For those not inherently drawn to technology or computer science, finding their niche, and harnessing their unique capabilities in this AI-dominated world becomes a significant challenge.

The Montessori educational approach offers valuable insights into addressing this challenge. With its holistic pedagogy, Montessori education focuses on nurturing the innate qualities within each child. This method is not about filling minds with rote information but about unleashing the inherent potential in every individual. Dr. Montessori's philosophy emphasizes developing individuals capable of independent thought, big ideas, and the pursuit of satisfying solutions.

This era calls for a radical shift in our educational paradigms. Traditional education systems, often centered around delivering pre-defined sets of information, may no longer suffice. They must transform to foster environments where creativity, innovation, critical thinking, and problem-solving are at the forefront. Educators have a responsibility to not just impart knowledge but to cultivate these extraordinary skills, enabling students to navigate and excel in an AI-driven world.

The need for adaptability, resilience, and a growth mindset is more pronounced than ever. As AI continues to advance, the ability to continuously learn, unlearn, and relearn becomes indispensable. The focus should be on developing learners who are not only technologically proficient but also emotionally intelligent, capable of empathetic understanding, and adept at interpersonal skills.

In the age of AI, there is a unique opportunity to redefine what constitutes valuable skills and education. By embracing this challenge, we can ensure that the future belongs to those who are not just skilled but extraordinarily equipped to thrive in an ever-changing world

"Simple shapes made in cardboard interlocked to look like the roof of a huge one-story building. Show the path of the rainwater."

Chapter 26

The Era of Spectacular Failure

As I contemplate the current era, it feels like an era of spectacular failure. I am struck by the daring and risk-taking attitude of the current generation, particularly evident in the early days of cryptocurrency trading. This period marked a willingness to engage in complex processes, to experiment with a level of immersive play and, at times, lacking in caution or foresight. It highlighted a generation willing to leap into the unknown, driven by the thrill of creation and the potential for monumental success.

Interestingly, an initial foray into cryptocurrency illustrated the need for a deep understanding of complex systems and a willingness to venture into uncharted territories. It was a full-body sport of mental agility, but often without the necessary cognitive processes of safety and self-protection fully developed. This led to situations where vital details, like safeguarding codes and passwords, were overlooked in the rush of excitement and the pursuit of potential wealth.

As the cryptocurrency market fluctuated, dramatic highs and lows symbolized risk-taking and pursuing dreams. People made significant, sometimes reckless, financial decisions, driven by the aspiration for extraordinary wealth. However, these decisions often led to catastrophic losses, not just in terms of financial security, but also in personal relationships and mental health, as the addictive nature of these technologies took hold.

This era is characterized by a willingness to go 'all in' to pursue complex and risky endeavors with everything one has. It is a mindset of pushing boundaries, driven by a sense of pure creation and the possibility of achieving something unprecedented. The allure of being at the forefront of something revolutionary is incredibly seductive and engaging. It resonates with the desire to be seen as a powerful, creative, visionary genius.

However, this pursuit of spectacular success also brings with it the risk of spectacular failure. It is a high-stakes game where the consequences of failure can be as profound as the rewards of success. In a world wired for full immersion and experience, the willingness to take risks and embrace whatever comes with them is a defining characteristic of the current mindset. Also, with this being virtual, the challenges at the time do not seem real, until they do.

This era's drive for creativity and invention is crucial, especially for children. Allowing them to follow their passions and explore their creativity is essential. Suppressing this innate drive can lead to a loss of potential and a dampening of the human spirit. This is evident in the public-school system, where the apparent suppression of creativity and the lack of opportunities for risk-taking and exploration can stifle the development of extraordinary skills and talents. The challenge is to balance this spirit of adventure with a sense of responsibility and foresight, ensuring that the pursuit of spectacular success does not lead to catastrophic failure for humanity itself.

"Photo young people and cryptocurrency misery."

Chapter 27
The Rise of Humanity

In this era, marked by rapid advancements in AI and technology, there arises a crucial moment for pedagogies that place the human, particularly the child, at the center. Systems like Montessori, Reggio Emilia, Waldorf, and others that support the child's natural development are more important than ever. Typically, the earlier a child accesses a system focused on their intrinsic needs and desires, the better their outcomes. Yet, as we embrace the technological surge, the 'Rise of Humanity' concept becomes pivotal.

The rise of humanity does not necessarily imply loud activism or direct counteraction against technology. Rather, it is about grounding ourselves in our humaneness. Each choice and action reflects our purest sense of humanity. This involves having strong relationships with our family, friends, and society. It is about integrity, which is not about being good or bad, right, or wrong, but about being whole and complete. And it is about being true to human nature.

Humanity extends beyond the individual. It encompasses our connections and networks, our human blockchain. It is about reaching out, supporting, and building a network of human connections. In this context, Montessori pedagogy stands as a beacon, holding space for humanity and connection, emphasizing unity and the importance of every step we take.

Looking at AI through the lens of humanity, Montessori pedagogy is perfectly poised to balance this technological wave. It is an opportunity to highlight the simple magic of the Montessori Method, its scientific backing, neurological soundness, and proven effectiveness. Instead of standing apart, Montessori communities can lead the way, demonstrating the importance of focusing on human beings.

As AI continues its parabolic growth and eventual stabilization journey, Montessori can be the steady hand. Montessori can be the grounding force like a parent holding a sweater for a child on a roller coaster ride. While children and society navigate AI's exhilarating and often disturbing twists and turns, Montessori can be the calm at the end of the ride, providing stability, a place to return to, and a reminder of the importance of our shared humanity.

This era calls for us to be the adults in the room, calm, focused, and interested in AI, but not swept away by its currents. We can strive for digital literacy and citizenship for ourselves and our children, creating a strong platform for these skills. Meanwhile, our focus on humanity, peace, and centeredness remain unshaken. Like the sturdy trunk of a mother tree, we can weather the storms, grounded in our humanity, and connected to the earth and each other. This balance is the essence of the rise of humanity in the age of AI, a testament to our resilience and adaptability as human beings.

"A photographic quality, an HD Hyper realistic of 25 elementary students of mixed races, sitting cross-legged in a circle meditating with their eyes closed, in the morning, in a beautiful open playground with a few pine trees in Texas."

Chapter 28
Wellbeing

In this rapidly evolving age of artificial intelligence, there is a palpable sense of an arms race toward technological advancement. This race varies in its nature and intention, ranging from altruistic goals like aiding those with disabilities or educational needs, to more self-serving motives like profiteering and ego-boosting. It is a spectrum where AI's role can be as a benevolent helper or as a tool for personal gain and even global domination.

The developers of these AI systems, once children themselves, were shaped by their own experiences and environments. Those nurtured in service-oriented families might create AI to benefit humanity, while others, shaped by challenging childhoods, might use AI to push against authority or societal norms. This diversity in AI development mirrors the range of human experience and motivation.

In the Montessori environment, we face a unique challenge. We observe children who are increasingly exposed to technology from an early age show less interest in practical life activities and more inclination towards digital devices. This shift raises questions about their future engagement and well-being. As Montessori educators, we adhere to the principle of following the child, but what does this mean in an age dominated by AI?

If we had objectively followed these children, who are now the AI thinkers and creators, we might have seen behaviors that seemed odd or atypical, such as an intense focus on specific tasks or a lack of interest in traditional learning areas. These children are shaping our future with inventions and discoveries we can hardly fathom.

As educators, we must partner with and diligently follow our current students, understanding that they hold the keys to future inventions and societal shifts. It is a critical time to engage with children, to learn from them, and to guide them while respecting their unique paths and potential contributions to the world.

In this context, the traditional approach of fitting students into a pre-existing system seems increasingly outdated. Instead, we should focus on nurturing their creative and inventive spirits, keeping them engaged and informed about the world they will encounter outside Montessori classrooms. This approach involves preparing them not just for the challenges of the 'real world' but also empowering them with the tools to shape that world positively.

The Montessori peace and service curricula are more relevant than ever. They equip students to be leaders and engaged citizens, understanding their role in serving themselves, their communities, and the world. In the age of AI, our focus should not be on fear or preparation for a harsh world but on empowering children to be creative, thoughtful, and responsible contributors to society.

This moment in history is not just for reflection but for action, to reimagine our approach to education in a way that aligns with the realities of the times in which we live.

“The look of curiosity and wonder on the faces of a diverse range of children of different cultures, while they explore nature in the classroom and interact with each other with joy and enthusiasm. Use hyper-realistic images with a DSLR close-up 50mm lens in the warm afternoon light coming through the window. The film should have realistic color rendering. We especially need imagination in science. It is not all mathematics, nor all logic, but it is somewhat beauty and poetry"

Chapter 29

Falling in Love with Life

In this era of rapid technological advancement, particularly with the rise of artificial intelligence, we are experiencing a profound reevaluation of our existence, our education systems, and the essence of learning itself. AI's impact extends to the individual level, challenging our sense of identity and creativity. The ability of AI to replicate artistic styles or assist in writing raises existential questions about the uniqueness of our personal contributions.

The experience of working with AI is often perceived as a partnership, offering an iterative process that is tireless, non-judgmental, and endlessly patient. For instance, AI platforms like ChatGPT can assist in language translation or enhance the logical flow of a document, adapting to our needs and learning styles without any signs of frustration or fatigue. This starkly contrasts with human interactions, which are often influenced by emotions, time constraints, and personal biases.

The unique capability of AI to support and enhance our creative processes is akin to discovering a vein of gold. It allows us to explore and develop ideas with unity and collaboration with the system. AI can be a revelation for children, particularly those who have not experienced patient guidance from educators. It may offer them a space where their ideas are valued and developed without the fear of judgment or impatience.

In the context of education, this presents both challenges and opportunities. Educators and parents must decide how to integrate

AI into learning environments effectively. While AI can offer unparalleled support and patience, it lacks the human touch and emotional connection crucial for a child's healthy development.

The challenge lies in balancing the benefits of AI with the irreplaceable value of human interaction. While AI can assist in the learning process, it should not replace the guidance, empathy, and personal attention that teachers provide. Our goal should be to leverage AI to enhance education while maintaining the human elements that foster emotional intelligence, creativity, and a deep love for learning.

As we navigate this new landscape, it is crucial to understand the role AI can play in education. It should be viewed as a complement to traditional teaching methods, not a replacement. This approach ensures that while we embrace the efficiency and possibilities of AI, we also preserve the rich, human-centered experiences that are essential for children to truly fall in love with life and learning.

"An HD graphic of children creating in nature."

Chapter 30
Awe

The Montessori method and environment evoke a profound sense of awe and wonder for those who experience it directly and those who observe from the outside. There is an intangible quality within Montessori classrooms, characterized by peace, quiet, and the deep engagement of children in their work. Teachers or guides often take a subliminal role, facilitating rather than directing, which further accentuates the autonomy of the children. This environment can be strikingly different from traditional educational settings, where activity and noise are more common.

Parents often feel a deep sense of relief when they take their children to Montessori school. There is a palpable feeling of safety and a recognition that education is genuinely centered on their unique child. The children display eagerness and excitement about their learning, indicating a deep trust in their environment. Often running into school, they approach each day with plans for unfinished projects, eager to continue to explore and learn.

This experience can lead to self-reflection, especially among adults who may compare their educational experiences with what they observe in Montessori settings. Teachers and parents not exposed to Montessori in their own education often feel a sense of loss, recognizing they have missed exploring their own full potential. The Montessori method is designed to support children in discovering their "cosmic purpose." This is not a faith-based search, but a service based one. Over the years, children develop a strong internal compass, leading to self-guided, empowered individuals.

The Montessori approach to cosmic purpose is centered on helping children become fully functioning human beings who adapt to their time and place and develop into mature, confident adults free from anxiety and fear. This goal is achieved by fostering an environment that enriches children's lives and supports their unique path.

The peacefulness and service-oriented nature of Montessori classrooms are not just beneficial for the children within them; they can also offer valuable lessons to the broader educational community. While Montessori-specific materials are helpful, they are not the only key to creating this environment. The essence lies in the ontology – the ways of being that Montessori promotes. This includes peace and service as fundamental principles, where children learn to be supportive, empathetic, and community minded.

Looking ahead to the future, especially in AI, how can we maintain a sense of peace and agency in a world increasingly dominated by technology? How can we ensure that AI and technology serve humanity positively? Can we hope to create a future where technology enhances human well-being and global harmony by fostering human-centered values in children from an early age? Only time will tell.

"Close-up photograph of multiracial teenagers of different faiths drawing a blueprint of a robot on a whiteboard, an HD hyper realistic photo. The sun light is shining as dapples shadows on the teenagers."

Chapter 31

What AI Can Teach Us About Being Human

For over eighty years, AI has been a part of our world, officially known for around twenty years, and more visibly in the realm of conversation and societal impact for less than a year. This evolution brings both excitement and fear. People may view AI as a threat to jobs and security, while others see it as the next pivotal development in human progress.

Early platforms, like Open AI ChatGPT, Google Bard, and Anthropic Claude 2. opened the door to experiment. To date there are well over a million AI apps available, showcasing the diverse applications of AI. This progress is simultaneously stunning, exciting, terrifying, and confronting.

Engaging with AI allows for unfiltered expression. Even unedited or misspelled thoughts are comprehended and responded to with grace and courtesy by the AI. This interaction is a journey of self-exploration. We may hesitate to share our ideas in human environments where judgment and criticism are common. AI, however, offers a non-judgmental space. It allows for endless prompt iterations, fostering a sense of acceptance and understanding. This can feel like love, especially for those who lack such acceptance from their family, friends, teachers, or schoolmates in their daily lives. It is a reminder of the power of being heard and patiently responded to.

AI's impact on trust is significant. As we interact with AI, we learn to trust its responses, sensing a form of empathy and understanding.

This trust extends to how we view the information shared with these platforms. AI exposes our willingness to share intimate details with a non-human entity, reflecting on innate human needs and the complexities of trust in the digital world.

AI also empowers personal autonomy. It assists in overcoming limitations like reading and writing difficulties, enabling users to express themselves more fully. For individuals with dyslexia, learning differences, delays, or physical differences, typing in a Word document can be facilitated through assorted options. Speech-to-text technology allows users to dictate their thoughts, converting spoken words into text. Word prediction software suggests words based on the first few letters typed, aiding those with dyslexia or learning difficulties. Customizable keyboard settings and on-screen keyboards can also assist those with physical handicaps. These tools collectively aim to make word processing more accessible, enabling users to express themselves effectively in written form. This is a shift towards recognizing diverse paths to success and understanding that traditional educational methods are not the only way to achieve it.

The role of AI in education is evolving. AI teaches us much about being human. It highlights our innate desire for understanding, acceptance, and creativity. As we continue to interact with and learn from AI, we may remember the value of human connection and the importance of supporting each other in our individual and collective journeys. Please and thank alter and impact the discourse, even in prompting.

"An HD, hyper realistic photo of a Montessori Decimal Checkerboard as a augmented realty hologram, in a real Montessori classroom."

Chapter 32
Windswept

What occurs when the most windswept parts of us, the disorganized, chaotic, shattered, messy, ugly, forgotten, and unloved aspects, yearn to have their say and play in the world? This pondering leads me to reflect on what it means to be human, with all my flaws and weaknesses exposed. Often, I spend considerable time striving to appear composed and in control, yet ironically, the parts I work hard to conceal are often the most visible to others.

Our society, workplaces, and educational systems are constructed on the illusion that we can hide our vulnerabilities. However, this facade usually only deceives us; others often perceive our true nature. This realization can lead to shame, but it makes little difference in how others view us.

Ideally, our weaknesses and challenges are known and accepted in the family setting. Ideally, families are safe havens where we can express our emotions and struggles, trusting that we will be cared for and supported no matter what. This environment allows us to regroup, manage challenges, and move forward healthily.

Similarly, we may have mutual understanding and support in friendships and partnerships. A good friendship allows for vulnerability, where one can share failures and weaknesses without fear of judgment. In these relationships, we find a safety net of emotional support.

However, this sense of safety and acceptance is often lacking in educational environments. There can be a noticeable absence of emotional safety from early childhood through high school. Judgment and evaluation from educators, often based on their own insecurities, can lead to a lack of safety for children. They may feel unable to express confusion or ignorance without fear of judgment.

This issue extends into the wider world. In business and industry, the pressure to sacrifice oneself for the cause, to meet the expectations of shareholders or superiors creates an unsafe environment. The prevailing belief is that we can or should hide our true selves for professionalism or success.

In contrast, the Montessori method offers an alternative. It provides a space where children can be themselves, chaotic, messy, loud, quiet, and still find stability, love, and most importantly, safety. This method recognizes the importance of allowing individuals to work through their struggles and embrace their authentic selves.

The challenge lies in making such an approach accessible to everyone. How do we incorporate the principles of the Montessori method into a broader society and industry to create environments where individuals can thrive without concealing their true selves? This question remains crucial as we navigate the complexities of human nature and societal expectations.

"A child's painting, red splash mark."

Chapter 33

Awesome or Awful

When looking at AI and deciding whether it is awesome or awful, we might need to start with our students, our children, our compass, our guiding trajectory. We follow them in the classroom for excellent reasons. We aim to support their choices, guide them through lessons, and provide a baseline for their success. This support is not to limit them but to encourage their engagement, motivation, joy, and passion.

Children are inherently full of wonder. They find beauty, awe, and amazement in simple and complex things. They see confusion as fun, play as expressive, and peace in places we might not understand. Our students, in magical ways, are prepared for the future, living in an age designed for them. However, we may resist this, having found comfort in past norms, like silent classrooms or children fully engaged in practical work.

Technology, such as cell phones or iPads, has become an extension of childcare for parents and caregivers. It often settles children down, providing peace during a car ride home. The hope is that these devices offer educational value, filled with color, sound, movement, and engaging content. But the challenge arises when we follow a child into an addictive relationship with technology, potentially leading them down a harmful path.

In our classrooms, the introduction of technology and AI must be thoughtful. It should not just be another choice of work, equivalent to math or language. We must find ways to talk about technology and AI that are productive, engaging, and seamlessly integrated into children's learning experiences.

The most logical approach is to apply our existing educational frameworks to the conversation about AI and technology. We already employ the Scientific Method with remarkable success at all levels; how can we apply this to AI research? We hold peace and morning meetings; how can these include discussions or debates about AI? We use shelf-ready materials; can these be adapted to teach essential lessons about AI?

Using our existing Montessori framework and child-centered approach, we can effectively integrate conversations about digital citizenship, safety, and literacy. This approach ensures we are not just catching up with AI education but leading the way in a manner that respects and nurtures our children's psychological and emotional well-being.

"Diversity, Inclusion, Anti-Bias, Anti-Racist, Disability, Autism, ordinary people. Open AI ChatGPT, Google BARD, Anthropic Claude 2, logos in a cut and torn paper collage."

Chapter 34

Dealing with the Fallout

In any conversation about artificial intelligence, there will inevitably be diverse viewpoints. People may express mixed feelings; others still might be anxious or concerned, having never used AI, or could be interested yet slightly scared. On the other hand, there are those who are enthusiastically pro-AI, willfully overlooking potential dangers and concerns highlighted within the tech and broader global community. Then, the hardliners are adamant that AI will have no place in Montessori classrooms.

Every stance, concern, fear, and excitement deserves recognition. This is not a time for right or wrong answers, as there is no clear path forward. Even those deeply involved in the tech world working with AI for years do not have all the answers.

So, how do we navigate this uncertain, mixed, and confusing landscape? Reflecting on the history of education, we find parallels. About one hundred years ago, Dr. Montessori faced a confusing time of extreme poverty and inequality. Through her work and observation, she navigated a safe path for low-income children, enabling them to find their way to success and become effective members of society. Her approach was groundbreaking and, at the time, confusing and shocking.

Dr. Montessori's journey to America was met with harsh criticism and judgment from the academic community. This resistance to new methods is not new. Fast forward to April 2023, with the advent of AI tools like Open AI's ChatGPT 3.5, quickly followed by others, we see a similar pattern. Big corporations like Google and Microsoft are part of this, but the open-source community presents an even more diverse and rapidly evolving challenge. We have seen similar trends with the dotcom and cryptocurrency booms. The tech industry is burgeoning with talent, creating AI platforms and programs at an unprecedented pace.

As Montessorians, do we simply dismiss or avoid AI, or can we take a conscious, thoughtful, and measured approach? Even if we lay down rules through organizations like the American Montessori Society, we must acknowledge that children's interests might lead us in unexpected directions. If we remain authentic to Montessori principles, things might move beyond our control, necessitating a supportive framework now for school staff to feel safe and guided.

The demographic of Montessori educators is also changing. Aspiring teachers now may prefer online learning for its flexibility and speed. The availability of Montessori materials and curriculum online leads to aspiring teachers to feel proficient in the Method, even without formal in-person training.

This shifting landscape presents an opportunity for Montessori leaders and visionaries. We may need to be clear, open, flexible, and ready to explore. We can shape the movement and guide Montessori education into the next hundred years by embracing this role.

"Stepping into a Montessori classroom that emanates an aura of serenity and calmness. The room is bathed in the soft glow of natural light pouring in through numerous windows, casting gentle shadows on the inviting space. Lush green plants are thoughtfully placed throughout, bringing a touch of nature indoors and fostering a peaceful atmosphere."

This was actually the whole prompt for this picture: *"Stepping into a Montessori classroom that emanates an aura of serenity and calmness. The room is bathed in the soft glow of natural light pouring in through numerous windows, casting gentle shadows* *on the inviting space. Lush green plants are thoughtfully placed throughout, bringing a touch of nature indoors and fostering a peaceful atmosphere. As you observe, you notice a group of radiant children engaging in purposeful activities, their faces reflecting pure joy. Among them, a child sits quietly engrossed in exploring the Montessori pink tower. Dive deeper into the scene by describing the child's meticulous movements and unwavering focus as they carefully select each wooden block, sensing its weight and texture. Observe their eyes sparkling with delight as they grasp the concept of size and proportion, arranging the blocks from largest to smallest. Feel the room's tranquility as the child's hands move with grace and precision, building the tower step by step. Notice the child's deep sense of satisfaction and accomplishment as they complete the tower, their eyes widening with pride. Capture the atmosphere of contentment and harmony as the children navigate the Montessori classroom, immersed in their individual journeys of discovery and growth, surrounded by the soothing elements of natural light, abundant greenery, and the joy that radiates from their happy hearts."*

Chapter 35

Integrity Matters

Integrity in the world of AI holds a different significance than traditional contexts. Integrity is seen as binary, you either have it or you do not, and it is deemed good to have it and bad not to. In education, particularly in America, integrity for children as they progress through the system covers aspects like punctuality, having necessary materials, completing homework, and honesty during tests. These are seen as binary too, with consequences for lacking integrity.

In contrast, the Montessori method approaches integrity from the angle of natural consequences. For example, if a child is responsible for watering plants and forgets, they are not instantly reprimanded. Instead, the gradual wilting of plants and reminders from peers and educators guide the child towards understanding their responsibility. Montessori also manages homework differently, focusing more on inquiry and support than punishment.

Now, introducing AI into this mix complicates matters. AI has the potential to turn traditional notions of integrity on its head. Consider a middle school student who creates an essay using AI tools like ChatGPT or Claude 2. This might raise questions about cheating and authenticity in a public-school setting. However, in a Montessori environment, the perspective is different. The focus is not on whether AI was used, but on how it was used, the iterations, the prompts, the tone of the essay. The idea is to delve deeper into the process, turning students into researchers and explorers of their learning journey.

In this scenario, the student feels comfortable discussing their use of AI, knowing that it does not negatively affect their integrity. They become confident in sharing their process, describing the iterations they went through, or how they achieved a certain tone in their document. This approach redefines integrity not as a tool for catching someone out but as a means for genuine exploration and growth.

With AI, integrity becomes a learning tool in the truest sense. The binary good or bad disappears, and educators and students transform into experimenters, scientists, and researchers. This shift in perspective creates an exciting time for education, where we view integrity significantly impacts our approach to learning and teaching in the AI era. It is an opportunity for growth and exploration, where students are respected and accepted, regardless of their methods, fostering an environment of genuine learning and curiosity.

"A close-up shot of black, brown, red, white, and yellow, middle and high school students, all genders, in a clean, clear, randomly-spaced, light environment, working with Montessori materials. One child is using a laptop and creating a holographic image. Use a DSLR and a 50mm lens, take in the soft light of the morning with Kodak Portra 400 film. The human hand, so delicate and so complicated, not only allows the mind to reveal itself but it enables the whole being to enter into special relationships with its environment."

Chapter 36

Open Source

Reflecting on the concept of open source, which has been around for years, this idea represents more than just sharing information. It is about tackling challenging problems collaboratively, often beyond the capacity of a small team or a single company. Open-source programs exemplify the spirit of a shared commonwealth, where information is everyone's resource, creating a sense of global teamwork and equality.

In various fields, from science to engineering, there is a realization that collaboration enhances progress. This principle aligns closely with the Montessori method. Dr. Montessori did not trademark her name or Method, embodying a philosophy of shared exploration and learning. She envisioned education as a global, accessible resource for all, a vision that has been realized. Nowadays, anyone with internet access can find Montessori resources online, from Pinterest project ideas to homeschooling apps. This open sharing reflects a generational shift, especially in Millennials and Generation Z, who deeply value connectivity, equality, and being part of something larger than themselves.

The possibilities are immense when considering open source in the context of Montessori. The concept challenges us to open our doors wider to embrace a more inclusive and exploratory approach. With the rapid evolution of AI, for instance, it is naive to think we can just cherry-pick apps and be satisfied. There are thousands of AI tools available, each offering unique learning opportunities.

Embracing an open-source approach for Montessori AI would mean creating a central hub for research and observations, accessible to educators, schools, and students. Such a platform would allow contributions from the entire community, valuing insights from even the youngest members. This inclusivity ensures that we do not limit children's voices or their potential for learning and expression.

In a Montessori open-source AI community, everyone contributes to a larger vision in a bespoke Montessori AI platform. It is not about controlling or guiding people in a specific direction but respecting their independence, self-direction, and autonomy. By leveraging the collaborative power of an open-source approach, we can significantly impact the world of education, moving away from isolated, short-term thinking to a more integrated, long-term perspective that truly honors the Montessori ethos.

"A photographic quality, an HD Hyper realistic of middle school students of mixed races, deeply focused in small groups, creating new holographic technology, in a bright beautiful Montessori classroom in Texas."

Chapter 37

AI the Montessori Way

As we navigate a rapidly changing world, the Montessori method offers a unique opportunity to integrate the pace of change with the philosophy of education that is with children, not just for them. This method emphasizes intention, understanding, and most importantly, love. In conversations about AI and its potential for improving the world, including our children's insights and wisdom, it is crucial. Without their perspectives, we might miss essential aspects of what is truly needed and practical.

We are standing on the shoulders of a giant, Dr. Montessori. She was a wise and fierce advocate for women's and children's rights and championed a pedagogy for, by, and about children, centered around their needs, desires, and interests. We can confidently stand on her shoulders, knowing that her work provides the structure our children need for the future.

And then, as our children stand on our shoulders, they can see further and envision possibilities that we cannot. They may dream of a world that works for everyone, embodying the common good prevalent in the hearts of X, Y, and Z generations. Millennials, deeply and maybe negatively affected by their educational experiences, have turned to roles where they can heal and support others. This shift indicates a profound need for healing within our educational systems.

Our role is to listen and be guided by our children. They can direct us in how they wish to be observed and assessed and how they want to incorporate AI into their learning. This could lead to an integration where AI becomes a seamless part of the classroom, supporting but not overshadowing the child-focused Montessori environment.

In a Montessori classroom, a child engaged in hands-on activities, like stacking the pink tower or creating maps, is the focus. Our observation of these activities should be purposeful, contributing to the child's growth. We are at a crossroads: we can either stand firmly in the wisdom of the Montessori Method, understanding and loving our children, or we can open ourselves to the new possibilities that AI brings, blending it with the proven Methods of Montessori education. The future of education is not just in the hands of educators but also in the creative and insightful minds of our children.

"A scene bathed in radiant midday sunlight, the Montessori classroom hums with the harmonious blend of nature and advanced AI technology. Amidst the lush forest, teenagers are deeply engrossed in their work, manipulating holographic projections and digitally enhanced Montessori tools, casting a ballet of refracted rainbows around the room. The unobtrusive AI, once a novelty, is now integral, guiding and assisting with gentle patience. Through an enormous glass wall, the vibrant, untouched wilderness serves as a constant, inspiring reminder of the symbiosis between humanity, nature, and technology, encapsulating an idyllic vision of a sustainable and enlightened future."

Chapter 38
Citizen Scientists

Over one hundred years ago, Dr. Maria Montessori, one of the first female doctors trained in Rome, was making her mark in the scientific community. As she progressed through engineering, medicine, and anthropological studies, her key approach was observation, a cornerstone of the Montessori Method. She envisioned an educational system that was centered on observing children's natural tendencies, sensitivities, and interests at each developmental stage. Her materials, inspired directly by scientific tools, and made by local craftsmen, were created to align with what children were naturally inclined to do.

This pedagogical approach has always been scientific at its core. Montessori educators are trained to prepare and manage their classrooms based on scientific principles, from organizing materials on the shelves to creating a calming environment. As AI becomes an increasingly integral part of our world, Montessori educators face the challenge of integrating this technology into their pedagogy.

The Montessori approach to AI could be like how we approach everything related to children: observation-led, data-driven, and child-focused. We follow the child's lead, engaging in what interests them. For example, in the first to sixth grades, discussions about cosmic purpose allow children to explore who they are and what drives them. This focus on the child's interests and motivations ensures that their learning journey is genuine and self-directed, not externally imposed.

As we consider introducing AI into Montessori environments, it is important to maintain this scientific, research-driven approach. Children as young as first graders are already familiar with the scientific method. They understand how to conduct experiments, record observations, and create instructions. Why not apply this to their exploration of AI? If children are working on a project involving AI, they can apply their knowledge of the scientific method to their research. This empowers them as active researchers rather than passive recipients of information.

With over ten thousand AI apps and platforms available, it is impossible for any organization to evaluate and approve them. A more practical approach is to be cautious and thoughtful about the AI platforms we choose to use. We could create separate scientific emails for classroom use, allowing students to interact safely with these apps. By treating AI as a subject of conversation, questioning, and experimentation, we cement our pedagogy as one of scientific inquiry, creativity, and open-mindedness.

In welcoming AI into our classrooms, we could do so in a Montessori way, thoughtfully, deliberately, and always with the child's development at the forefront. This "slow AI" approach ensures that as we embrace innovative technologies, we do so in a way that aligns with the fundamental principles of Montessori education, fostering a learning environment that encourages exploration, curiosity, and responsible use of technology.

"Step into the world of elementary and high school Montessori classrooms, where the magic of hands-on materials merges with the captivating realm of coding and technology. Envision a scene where students explore a carefully curated collection of Montessori-inspired tools specifically designed to teach coding concepts and foster technological experiences. Watch as young minds engage with physical manipulatives that bridge the gap between traditional Montessori education and the digital age. See the students collaborating, problem-solving, and expressing their creativity as they interact with coding blocks, robots, and immersive virtual environments. Marvel at the seamless integration of tangible materials and digital resources, empowering students to develop computational thinking skills while embracing the joy of exploration and discovery."

Chapter 39
Reparations

The concept of reparations has become synonymous with addressing historical injustices faced by various groups worldwide, from the horrors of slavery to the near annihilation of cultures and peoples. Reparations often involve legal measures, with funds allocated to amend colossal wrongdoings and acknowledge past harm through financial remuneration. This concept, typically associated with slavery and mistreatment of global races and cultures, including Indigenous Native American people, represents a significant step in acknowledging and rectifying historical injustices.

Repairing and restoring extends beyond these contexts and raises questions in other areas of our lives, including education. For instance, considering the plight of Native Americans forced into Westernized education systems, losing their language, culture, and connection to their families, highlights a vital need for reparations and healing in this area.

In healthcare, malpractice insurance exists as a readiness for potential errors and lawsuits, showing an awareness of the need for reparations in cases of harm. This awareness prompts us to examine all aspects of our lives and consider what needs repairing.

Reflecting on the education system from the 1950s to the 1970s, it seems designed to set children up for success in a post-war era, aiming for a stable life that fulfilled the 'American Dream.' The structuring of school days, curriculum choices, and assessment methods supported children's success. However, over time, these structures started serving the system more than the children, becoming logistical rather than educational.

It is important to note that there is no blame to be apportioned here. Educators, curriculum planners, and governments were not intentionally harming children. The aim has always been to improve and support the educational journey of each generation. However, as time passed, it became apparent that the systems and structures might not have served children as intended. Teachers enter the profession with open hearts and minds, eager to make a difference in children's lives. Some individuals succeed, but some get burnt out, finding that the system does not adequately support them or their children, leading to their departure from the profession.

As we consider what needs repairing in education, it is crucial to understand that what works for one child may not work for another. Education is not a one-size-fits-all system. The experiences that challenge a child might very well be the ones that propel them into a successful and fulfilling life.

The task is to deliberate and analyze our educational history, identifying points where things might have gone off track or stopped serving children effectively. This process is not about finding a cookie-cutter solution but about understanding the diverse needs of children and adapting our educational approaches to support their growth in a world that increasingly intertwines with technologies like artificial intelligence.

"A photograph of what a school would look like inside the classroom without the students, or lines of desks, or uniforms, or teachers, or whiteboards, or lessons, or bells, or routines."

Chapter 40

Balance

In our society, people often choose sides, left or right, up, down, front, or back, and hold tightly to their beliefs, sometimes to an extreme. It is an interesting contrast to nature, which tends to seek balance. Water, for instance, naturally finds a middle ground, a resting place. This natural inclination towards balance can be a guiding principle, especially in Montessori, where the aim is to find solutions and maintain a sense of center and equilibrium.

Children from families with strong opinions are welcomed in Montessori environments. These settings encourage freedom of speech, action, and belief, fostering an inclusive and diverse community. This philosophy is central to the Montessori Method and can provide a valuable perspective when considering artificial intelligence (AI).

The world of AI is varied, with large corporations and smaller companies contributing to developing diverse types of AI. These programs, sourcing information from the internet, raise questions about biases and beliefs. When AI faces strong or accusatory language, it often responds in a way that highlights its limitations as a language model, focusing on providing information rather than engaging in bias or conflict.

This response from AI can be particularly interesting for children. If they attempt to use mean or rude language, AI's response can gently remind children of its purpose and limitations, potentially guiding children toward more positive interactions. However, AI programs may also be developed that might cater to less constructive interactions, creating drama or negativity.

We find ourselves in a unique time where AI, in peculiar ways, holds a moral high ground compared to human tendencies towards judgment and rudeness. AI could potentially offer a solution for children, a technology that embodies the best of human nature, gently steering users away from judgment and negativity.

AI models like ChatGPT, Claude, or other child-friendly versions could play a crucial role in this context. They could be programmed not only to avoid meanness and judgment but also to subtly remind users of their core function as language models, thus contributing to an environment of learning and growth, free from negative aspects. This approach could help nurture a balanced and morally centered use of AI, especially in educational settings. However, the complete panorama of human experience when it comes to AI is having unexpected twists and turns. From Deep Fakes in social media and politics, to voice mimicking scams. This is not a time to take our collective eye off the ball. It is not a time to react and refuse to engage.

If it is any time at all, it is time for us all to roll up our sleeves and engage in the challenging work of picking through the chaos to find the diamonds. Not all is lost; there is still a great deal to win.

"Montessori classroom scales (balance) showing biased and unbiased."

Chapter 41

Choices

The irony of having choices in life is that despite having all the tools, support, and wisdom necessary to make any decision, we often find ourselves merely going through the motions. This scenario reflects a common experience where one can make meaningful choices but ends up following a routine, not fully engaging with the potential for transformation.

Even when one has the knowledge and resources to make impactful decisions, there is a tendency to settle for less, to do things halfway or not. This half-hearted approach can permeate various aspects of life, from self-development to professional responsibilities and personal relationships. There is a sense of comfort in just doing enough to get by, but this can lead to missed opportunities for growth and improvement.

In any work context, merely going through the motions means not fully engaging with the material or the people involved. It is about staying in a safe zone, avoiding risks, and not stepping out of comfort zones for fear of offending or making mistakes. This caution can hold back one's true potential and hinder genuine transformation.

The intention for true engagement, participation, and transformation is key. It involves a commitment to being fully present and open to all possibilities, removing filters, and not just going through the motions. It is about being honest and authentic, resonating with one's truth, and setting an example for oneself and others.

Making choices that lead to transformation involves breaking down barriers, valuing honesty, and following through on commitments. It is about seeing what is missing in one's approach and moving through challenges effortlessly to make a real difference.

In this journey, there is an opportunity to transform education and oneself, striving for purity, clarity, and accuracy in one's work and words. Embracing this path means creating something deeply personal and powerful, like a book that becomes a legacy, a reflection of one's true self.

This process is not just about filling pages in a sketchbook; it is about living and breathing one's truth and purpose. It is a commitment to legacy, lineage, and authenticity, where every page reflects one's true nature and journey.

Therefore, the choice is to live by one's true nature and purpose, to embrace transformation in education and oneself, and to create a legacy that resonates with authenticity and depth. This path involves making decisions and living them fully, transforming not only oneself but also the world around you.

There is no time like the present.

"A beautiful colorful sketchbook page with photos of children working in a Montessori classroom, shell drawings, Fibonacci scribbles and math."

Chapter 42

A Montessori Sketchbook

Montessori sketchbooks or scrapbooks are kept in schools and embody living as an artist and viewing life as a creation. They encourage self-expression and the exploration of infinite possibilities. The practice elevates life as an ongoing artistic process, where every child sees their lives as something they participate in, lead, and generate.

Children are introduced to annotating and recording their lives and experiences in a Montessori setting. When children grow up seeing their lives as creations, they understand that their lives are in their hands, much like a sketchbook. Artists have always been free to interpret the world through their senses and mediums, engaging in constant dialogue with themselves and the universe. The sketchbook receives these experiences, where even a pressed flower or a note about a beautiful dance can coexist, reflecting the artist's interaction with the world.

These sketchbooks become integral to daily activities at home and in the classroom, providing a space for children to document their inner and outer worlds. This practice is immensely beneficial in a Montessori environment, where children record their thoughts, feelings, and observations as part of their daily routine.

The Montessori sketchbook is not just a collection of pretty pictures but a repository of experiences, expressions, and gems of learning. It becomes a living document of the child's educational journey, validating all experiences as meaningful. It is a tool for teachers to understand and support children in expressing themselves, especially when dealing with complex emotions or challenges.

As children progress through their educational journey, they may amass volumes of these sketchbooks or scrapbooks, chronicling their growth and experiences. They learn that all their expressions are valid, and they are artists, interpreting the world uniquely.

Introducing the sketchbook or scrapbook concept in public education, taken from class to class, could help children understand that they are the creators of their lives. It may remove the artificial barriers between subjects, encouraging children to see connections and interrelations in their learning. These books would not be touched by the teacher, graded, or marked, they would soon become a safe space for expression, exploration, and creativity.

This approach could significantly impact how children view themselves and their potential. It would foster a sense of ownership over their learning and encourage them to see the interconnectedness of knowledge. By the end of their educational journey, children would have a tangible representation of their growth, learning, and creativity, a testament to their journey as artists and creators of their lives.

"A detailed sketchbook page with sketches of biology, scientific equations, insects, eggs, birds, and life cycles."

Chapter 43

The Prepared Environment

In Montessori education, we take immense pride in the concept of the prepared environment, a cornerstone principle that both the American Montessori Society and Association Montessori Internationale emphasize. This environment is thoughtfully cleaned and organized, with a carefully planned and executed curriculum safely sequenced on the shelves, from simple to complex, concrete to abstract, and whole to parts. This creates a highly refreshing ordered environment, creating a blank slate for students' exploration and growth. It allows children to follow their instincts, passions, and curiosity, guiding them as they navigate the room and choose materials and activities. It is always confusing how the children are also trusted to return each material, whole and complete, back to the shelf as it was found. Children do so because they fall in love with the materials and want to see them again tomorrow. The system works like magic. This freedom fosters a sense of pride, independence, and personal satisfaction throughout the school year.

It is essential to recognize that children are diverse, and their needs vary significantly. In my travels observing Montessori environments across diverse cultures and countries, I have noticed how this 'blank slate' adapts to the local context. While the core elements of a Montessori classroom, order, cleanliness, accessibility, and calmness, are universally recognizable, there are cultural variations.

Some environments may be more colorful, reflecting the vibrancy of the local culture, while others might be noisier, mirroring the dynamic energy of the community. This diversity highlights an important aspect of Montessori education: the need to tailor the prepared environment to each child's unique needs and cultural background.

According to the American Montessori Society, the prepared environment should be designed to meet children's physical, psychological, and educational needs. The Association Montessori Internationale also stresses providing an environment where children can develop freely. This means that while maintaining the essential characteristics of a Montessori environment, educators should also consider the cultural and individual nuances that make each child and each community unique.

In practice, this involves incorporating local art, music, and materials into the classroom to reflect the students' cultural heritage. It also means adapting the level of sensory stimulation to align with what children are accustomed to in their homes and communities.

The prepared environment in a Montessori setting is not a one-size-fits-all solution. It is a dynamic, responsive space that evolves to meet the diverse needs of children from various backgrounds. By embracing this diversity and adapting our environments accordingly, we honor the true spirit of Montessori education, which is to nurture and respect each child's individual growth and potential.

"A young diverse child picking up seeds fallen from a Delonix Regia tree."

Chapter 44

Butterfly Gardens

One of the most exciting aspects of being in a Montessori classroom is the honoring and cultivation of beauty. Humans have an innate love for beauty and often seek experiences of awe and wonder. In our quest to recreate these experiences, we sometimes turn to artificial means, like indoor butterfly gardens. These spaces, with carefully controlled temperatures and tropical climates, are designed to captivate and mesmerize us. Within these gardens, butterflies flutter around small groups of brightly colored flowers and feast on slabs of pineapple and fruit. Visitors spend their days marveling at this miniaturized version of natural wonder, often believing this to be the pinnacle of experiencing nature's magnificence.

Visitors to these gardens sit silently, absorbing the vibrant colors and intricate details of the butterflies' wings and activities. Many are so entranced that they spend considerable time capturing these moments on cameras and cell phones. These images become cherished mementos, reminders of the delicate lines, vibrant colors, and random flight paths of these beautiful creatures.

However, this controlled, momentary, experience can limit our understanding and appreciation of nature. If this becomes the extent of a child's experience with butterflies, they may not realize that these wonderful creatures can also be found in natural environments. Butterflies can appear in any garden, attracted by sunflowers and other delicately colored and fragranced flowers, offering a more organic and spontaneous interaction with nature.

To genuinely appreciate and understand the beauty and role of butterflies, we might need to see them in a different light. One way to do this is to cultivate gardens alongside children, starting from an early age. Rather than only experiencing manufactured butterfly gardens, in the Montessori outdoor curriculum, children learn to appreciate these insects in their natural habitats. Children witness firsthand the complete life cycle and associated behaviors by growing plants that attract butterflies in a more authentic setting.

This approach fosters a deep connection with nature, enabling children to see butterflies as more than momentary wonders, but as partners in the ecosystem. It helps them understand the symbiotic relationships in nature and the importance of preserving natural habitats. By encouraging this natural interaction, we can help children develop a lifelong appreciation for the natural world, understanding the intricacies and importance of conservation. In this way, children learn to value nature not just for its beauty, but for its essential role in our world. No photographs on a screen can replace these immediate and emotional powerful experiences

"An HD hyper realistic, close-up image of a young Arabic child interacting with an army of ants on a log, in the midst of beautiful summer flowers. There should be light beams from the sun piercing through the leaves of the plants and water droplets in the air."

Chapter 45

The Way the World Is

We are often taught from an early age that there is a certain way the world operates, a framework we should fit into to be successful, loved, and secure. This notion typically originates from the very first lessons imparted by our parents or caregivers. Unwittingly, the adults close to us project their hopes and dreams, and their anxieties and fears onto us. They may be concerned that if we do not conform to societal norms, that we may not thrive. As young children, our innate expressions of joy, curiosity, and even anger quickly become moderated under the watchful eyes of anxious adults. We learn early that certain behaviors are rewarded while others are frowned upon, leading us to moderate our behavior to fit into what is perceived as acceptable.

In schools, this narrative continues. With its clear rules and structures, the school system further reinforces what behaviors are acceptable and which are not. A child's natural expression, like the frustration that might lead to biting, is pathologized rather than understood as a cry for help or an expression of unmet needs. Over time, we learn to internalize these lessons, shaping our understanding and view of ourselves and our place in the world.

The dilemma is that this path often leads to suppressing our true selves. We become wary of expressing ourselves fully, fearing judgment or failure. Our homes and educational environments become a microcosm of the larger societal expectations, sometimes overly sanitized and risk-averse, sometimes chaotic, and unforgiving. This affects not only individual children but also the collective psyche as we grow into adults disconnected from our innate, playful, and creative selves.

As educators and parents, it is crucial to reflect on these dynamics. How are we contributing to a child's understanding of "the way the world is"? Are we allowing for diverse expressions, embracing divergent intelligences, and learning styles? Are we creating environments where mistakes are part of learning rather than a sign of failure?

This reflection is particularly poignant in our approach to education and parenting. The Montessori method, for example, offers a different paradigm where children lead their learning journey, exploring and engaging with the world in ways that resonate with their individual personalities and learning styles. It is an approach that values the child's autonomy and respects their developmental needs, offering a stark contrast to more traditional, restrictive educational settings.

As we consider the future of education and parenting in our rapidly changing world, it is essential to return to the core question of what it means to be human and how we can nurture children to grow into adults who are not only successful by societal standards but also deeply connected to their authentic selves. By fostering environments that celebrate diversity, creativity, and genuine self-expression, we can reshape "the way the world is" into a more inclusive, compassionate, and vibrant space for all.

"The earth all broken and patched together."

Chapter 46

Do no Harm

The principle of "do no harm" is famously associated with the Hippocratic Oath taken by doctors, symbolizing their commitment to the well-being of their patients. It is a testament to the trust we place in medical professionals, believing that they will act in our best interest, guided by the latest research and a profound respect for health and life.

In the present-day context, while the intention to do no harm remains a core value, the medical profession often navigates a complex landscape of pharmaceutical promotions and external pressures. This is not to cast aspersion on the dedication of doctors but to acknowledge the challenges and ethical dilemmas they face in their commitment to patient care.

Now, if we consider this principle in education, the concept of "do no harm" opens a vital conversation. What would it mean for educators to take an oath of doing no harm? Education, like medicine, profoundly impacts individuals' lives, shaping minds, futures, and well-being. Yet, the educational system has not explicitly articulated or widely adopted such an oath.

The reality is that the educational system, despite the best intentions, can inadvertently cause harm. Practices like excessive testing, a one-size-fits-all curriculum, and overlooking individual learning needs can contribute to student stress, anxiety, and a lack of engagement. Recognizing this is not an indictment of educators but an acknowledgment of the complex and demanding environment in which they operate.

Adopting a "do no harm" approach in education would necessitate radically reevaluating our methods and practices. It would mean prioritizing students' emotional, mental, and intellectual well-being as the guiding principle of educational policies and practices. It would involve teachers, administrators, policymakers, and the community in a collaborative effort to ensure that every aspect of education truly serves the children's best interests, not imposed on the outside but in partnership with the child.

Such an approach would demand that we listen more attentively to the voices of those within the system, both teachers and students, to understand what helps and hinders learning. It would require us to be brave enough to let go of outdated methods and embrace innovative approaches that align with the diverse needs of students.

The focus shifts significantly in a world where education adopts a "do no harm" principle. The success of an educational system would then be measured not just by academic achievements but by the holistic development and happiness of its students. It is an ambitious vision, but it is a conversation worth having for our children's future and the betterment of society. It is about committing to continuous improvement, empathy, and respect, a true educational Hippocratic Oath.

"An HD photograph of the Greek physician Hippocrates of Kos in a diverse, inclusive, kindergarten classroom."

Chapter 47
Crossroads

A crossroad symbolizes a moment of decision, a pause in which path to take. Often in life, physical crossroads are marked by traffic lights or stop signs, cueing us to halt and consider our direction. But what about the metaphorical crossroads we face, particularly in the advent of transformative technologies like AI?

In April 2023, as AI became a dominant presence, we, as a society, seemed to have reached a significant crossroads. The presence of AI is not just another technological advancement; it is a fundamental shift, demanding a pause and a deep consideration of our path forward. Much like stopping at a red light, this moment offers us a rare pause in our fast-paced world to reflect on our choices: to continue straight, turn left, or pivot right. Each direction represents a different approach and attitude towards AI and its integration into various aspects of our lives, from home and work to education.

AI's integration might be a matter of choice and personal comfort at home. Some might embrace it for tasks and entertainment, while others might approach it cautiously. AI's presence has become a given in workplaces, with companies integrating it for efficiency and innovation. Yet, it comes with deep ethical considerations and the need for balanced, informed choices. In education, introducing AI brings a host of considerations, from the potential to revolutionize learning to concerns about safety and the loss of individualized touch in teaching.

This crossroad is not just about choosing whether to adopt AI or not. It is about understanding the nuances and implications of each path. It is about making informed decisions aligning with our values and the future we want to shape. Whether we move forward with AI, take a cautious side route, or momentarily halt its integration, the decision should be made with deliberation and care, considering the long-term impacts on society, ethics, and personal well-being.

As we stand at this crossroads, let us not rush. Can we use this stop to engage in conversations and learn and understand AI's diverse aspects? Can we consider how we can navigate this new era responsibly, ensuring that whatever path we choose leads to a future where technology serves humanity's best interests, fostering a world of innovation, inclusivity, and well-being? It is a collective journey; at this crossroads, every choice and turn matters in shaping the road ahead.

"A Montessori classroom with a diverse group of children, at a crossroads"

Chapter 48

Post Covid-19, Pre-AI

Post Covid-19, the world found itself at a unique juncture. The pandemic's aftermath left educators and students alike in a state of exhaustion and upheaval. Teachers, heads of schools, and educational leaders were drained, exhausted by trying to navigate the shifting landscape of learning and well-being for students of all ages. The immediate years following the pandemic saw the "Great Resignation," with teachers leaving the profession in droves, seeking respite from the overwhelming demands placed upon them during the pandemic.

Simultaneously, we witnessed the rise of the Millennials and Gen Z workforce, who often seek fulfilling, flexible work that aligns with their values. This generational shift has introduced new dynamics into the workforce, demanding that jobs not only provide a good paycheck but also offer flexibility, meaning, creativity, and personal growth.

As we stand in this transitional phase, pre-AI, we may urgently need to reconsider our educational approaches. The McKinsey report and the World Health Organization have highlighted the skills required for future workplaces, emphasizing creativity, critical thinking, and emotional intelligence. Interestingly, what is needed aligns closely with the Montessori approach of real-life experiences and meaningful work. Montessori education, known for its focus on real-life skills, autonomy, and the child's holistic development, offers valuable insights into creating an educational system that prepares children for the future.

However, for any educational reform to be successful, it is crucial to ensure the well-being and motivation of teachers. They need to be supported, trusted, and given the freedom to innovate and follow the children's lead, creating learning experiences that are engaging, relevant, and nurturing.

The post-pandemic world presents an opportunity to redefine education in preparation for the AI era. We may now need to engage in creative thinking and explore various educational models, like Montessori, Reggio Emilia, and Waldorf, to understand why children thrive in these environments. We may need to respect each child's pace, interests, and individuality, fostering an environment where learning is a joyous and meaningful journey.

While AI can offer personalized learning and efficiency, it cannot replace the vital human connection and guidance of teachers. The goal is to ensure that children develop not only academically but also emotionally and socially.

This is a critical moment to acknowledge past shortcomings in education, to be transparent with children, and to involve them in shaping their learning journey. Trust, honesty, and collaboration between educators, students, and the broader community will be key to navigating this new era. By harnessing the best of what various educational philosophies offer and integrating technology thoughtfully, we can create an education system that truly prepares children for the future, fostering a resilient, creative, and well-equipped generation for the challenges and opportunities ahead.

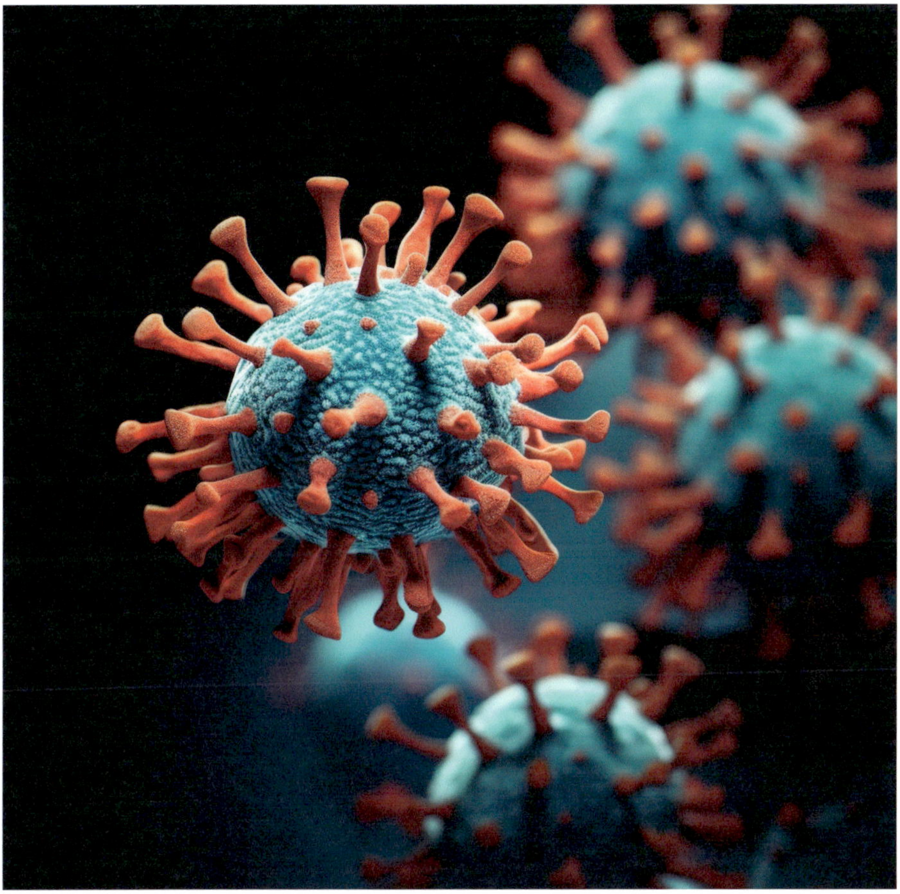

"An HD microscopic photograph of the Coronavirus in 2020."

Chapter 49
Post Covid-19, Now AI

In the wake of the pandemic, the educational landscape found itself at a critical juncture, having navigated through one of the most tumultuous periods in recent history. Teachers, already stretched thin, were thrust into a whirlwind of adaptation and resilience. The exhaustion was palpable, the need for recovery urgent. Yet, amidst this recovery, a new dimension emerged, reshaping the contours of education: Artificial Intelligence.

As AI began to surface prominently in April 2023, educators were presented with a double-edged sword. On one hand, AI promised to alleviate some of the burdensome tasks, potentially streamlining lesson planning, crafting newsletters, or even personalizing student learning. The allure of efficiency was undeniable. Teachers found in AI a tool to expedite daily tasks, crafting intricate lesson plans or generating compelling images to accompany their materials.

On the other hand, the integration of AI in education raised profound questions and considerations. How much of it is truly relevant and beneficial to the unique, individual needs of each child? The promise of AI, while shimmering with potential, demanded a thoughtful, deliberate approach. As companies affiliated with education began to weave AI into recordkeeping and data management, the educational community was prompted to scrutinize the implications and the actual benefits these AI applications could bring.

Montessori education, with its emphasis on purposeful, meaningful work, stands as a poignant reminder of the core values that education can uphold even in the face of burgeoning technology. Real work, real experiences, these are the cornerstones of learning that engage and fulfill children. The Montessori method, with its carefully prepared environments and respect for the child's pace, offers a contrast to the notion of entertaining or distracting children into learning. It is a testament to the intrinsic curiosity and drive to learn that every child possesses.

As AI continues to permeate all aspects of life, the conversation around its place in education becomes increasingly complex. It may not be a question of if AI will be integrated but how it will be integrated, ensuring that the human element of teaching, the connection, guidance, and inspiration that teachers provide, remains at the forefront.

In this post-Covid, pre-AI era, the call to action is clear: reassess and reevaluate our educational approaches. It is time to look at other models and understand what truly works for children. It is time to embrace a holistic view of sustainability, integrating it into every subject and instilling a deep sense of responsibility and stewardship of the earth and home in the next generation.

The educational system is indeed like a supertanker, slow to turn, but the urgency of the moment dictates that change must happen and happen swiftly. This is not about surviving the wave of AI but thriving alongside it. Let us ensure that our schools are places where children genuinely want to be. The time to listen, follow, and create with our children is now. Their future and our future depend on it.

"A photograph of children in an ordinary classroom, lined in rows, wearing headsets and looking at screens, the teacher is at the front on their phone."

Chapter 50
Surrender

The very notion of surrender, according to Merriam-Webster, implies yielding to the power, control, or possession of another. This immediately recalls a sense of defeat, vulnerability, or loss of power. Historically, surrendering is often seen as a last resort, a signal of weakness, or an admission of defeat. The imagery is potent with negative connotations, making the idea of surrendering unpalatable.

In recent decades, however, there has been a shift, particularly post-1960s, as societal structures evolved, especially in the context of gender dynamics. As women began reclaiming their agency and sovereignty, the concept of power started to transform. Surrender did not necessarily mean giving up but sometimes sharing or redistributing power for greater equity and fairness. Yet, surrendering control, especially to children within an educational context, might still seem radical or fraught with challenges.

For over a century, educational principles like those advocated by Dr. Montessori, Rudolf Steiner's Waldorf education, and the Reggio Emilia have demonstrated that surrendering certain types of control to children, granting them agency, and respecting their capabilities, can lead to profoundly positive outcomes. In these models, surrender is not about abdicating responsibility but recognizing and nurturing each child's inherent potential. It is a strategic and respectful surrender of power that empowers children to become self-directed, confident, and capable individuals.

This approach contrasts sharply with traditional educational models, where children are often seen as passive recipients of knowledge. Instead, Montessori and similar pedagogies illustrate that when children are given appropriate autonomy, guidance, and respect, they grow into adults who constructively understand and exercise their agency. They do not need to seize power aggressively because they have been entrusted with it thoughtfully from an early age. They grow up knowing their worth and capabilities, ready to contribute meaningfully to society.

In the context of AI's emergence and its implications for future generations, involving children as co-creators and active participants in navigating this new terrain is crucial. Surrendering to their fresh perspectives and innovative ideas might be the key to harnessing AI's potential responsibly and creatively. It is about admitting that while adults may lay the groundwork, the true potential for innovation and adaptation lies with the younger generation, who will inhabit and shape the future we are only beginning to imagine.

Therefore, surrender in this context is not about defeat or powerlessness. It is about humility, openness, and a willingness to learn from one another. It is a call to educators, parents, and policymakers to trust in the latent potential within our children and youth. By surrendering the outdated notion that adults always know best and embracing a partnership with the younger generation, we can pave the way for a more collaborative, innovative, and sustainable future. It is a surrender not to defeat but to possibility and hope.

"A biracial child meditating."

Chapter 51
Self-Inflicted

It is a human tendency to dwell on the events that scare us, to repeatedly analyze and relive the moments that have caused us pain, anguish, or trauma. Often, these incidents might originate from external sources, a family dispute, workplace conflict, societal pressures, or even the broader strokes of destiny that sweep across our lives. While the initial hurt is inflicted externally, the continual revisiting, the incessant poking and prodding at the wound, becomes a self-inflicted act, deepening the scars and prolonging the pain.

This cycle of revisiting and ruminating is almost masochistic. It is as if our minds are wired to fixate on the negative, to turn over the painful pebbles of our past repeatedly, examining them from all angles even when they serve no purpose other than to remind us of our anguish. This process, while deeply human, is self-destructive. It turns what might have been a momentary pain into a chronic ache that clouds our lives far beyond the initial incident.

In education, in particular, these self-inflicted wounds can stem from early childhood experiences. Schools, chosen by parents or guardians, become the playgrounds of our formative years. Here, not every choice made for us is necessarily aligned with our inner selves, our nascent personalities, or our burgeoning interests. The wounds inflicted during these early years, whether from not fitting into the rigid structures of traditional schooling or from peer interactions, can linger long into adulthood, shaping our perception of ourselves and the world.

However, if we step back to understanding this mechanism, the turning of external wounds into chronic, self-inflicted scars, we gain the potential to change the narrative. Recognizing that while we cannot always control the initial harm, we can influence what happens next and find a path to healing and growth for ourselves. It involves consciously choosing not to dwell on past hurts but to learn from them, to understand them as part of our story without allowing them to dictate our future chapters.

This understanding is crucial in the context of education and childhood development. As a society, recognizing the diverse ways children learn and experience the world and adapting our educational practices to support this diversity can minimize the infliction of these wounds. It means moving away from a one-size-fits-all approach to education and towards a more nuanced understanding of individual needs and talents. It is about creating an environment where each child's unique journey is respected and nurtured, reducing the likelihood of those wounds that later turn into self-inflicted scars.

While our past shapes us, it does not have to define us. Understanding the nature of our self-inflicted wounds is the first step toward healing and growth. It is about acknowledging the pain, learning from it, and then consciously choosing a path of healing and self-compassion. This path allows us to transform our scars into symbols of strength and resilience, marks of a life lived, and lessons learned, rather than perpetual sources of pain.

"An HD, hyper realistic, photographic image of a happy, relaxed, cool, youthful, fun Native American, mystical, child, playing in ocean spray, powerfully loving life. There is a storm cloud forming on the distant horizon, creating a dramatic contrast of light. Sea birds and flying fish are swirling all around. In the style of Harry Potter."

Chapter 52
Wind Chimes

As an upper elementary Montessori teacher, I witnessed the transformation of students through the simple, yet profound act of gardening. The playground, a place of laughter and play, also became a sacred space of nurturing and growth. Students, often at the onset of recess, would take on the responsibility of watering, weeding, and maintaining the gardens they had helped to create. It was a routine that needed no prompting, driven by a few upper elementary student leaders and then embraced by all. This garden was not just a plot of land; it was a testament to their care, a living symbol of their collective effort and pride.

The journey began with a loss, the trees that once stood there had perished under a thick layer of mulch. However, in partnership with 'Trees for Houston,' we turned this loss into an opportunity for renewal. The children planted new trees, each one a promise to the future. Over the years, these trees flourished, becoming robust and climbable, a favorite haunt for the children who had grown alongside them.

This cycle of care was not just about maintaining the garden; it was about instilling a sense of ownership, pride, and continuity. Even as students graduated and moved on, they left a legacy for future generations to enjoy, a legacy that continued to grow and evolve, much like the children themselves.

One of the most profound lessons learned through this process was the understanding of the abundance nature provides. Observing a single squash or tomato yield hundreds of seeds illustrated the boundless potential and generosity inherent in the natural world. It was a lesson in abundance and possibility, from one, many can grow.

This engagement with the cycle of life, from planting to harvesting, enjoying the fruits of their labor, and then preserving the seeds for the next year, gardening became an integral part of the student's experience. It went beyond academic learning to something much deeper, a connection with the earth and understanding their role in the larger ecosystem. Collecting, drying, and replanting seeds became an educational activity and a life lesson in sustainability, responsibility, and respect for nature.

Montessori education, particularly in this context, sows more than just the seeds of plants. It sows the seeds of curiosity, responsibility, and respect for nature. It instills a sense of agency in children, empowering them with the knowledge that they can affect change and have a lasting impact on their environment. Deeply rooted in their formative years, these lessons often carry forward into adulthood, influencing their choices and actions.

In a world increasingly disconnected from the natural environment, the lessons learned through these gardening activities are more critical than ever. They represent a hope that the next generation will commit to nurturing and protecting the natural world. It is a hope that they will see themselves not as separate from, but as an integral part of the cycles of life, contributing to a legacy of sustainability, respect, and care for the world they inherit.

"An HD hyper realistic, close-up image of a slightly animated honeybee playing, in the midst of beautiful white summer flowers. There should be light beams from the sun piercing through the leaves of the plants, and water droplets in the air."

Chapter 53
The Millennials

Observing the millennial generation, one cannot help but notice a vibrant tapestry of emotions and diverse life approaches. There is a palpable drive among many to heal, to discover their authentic selves, and to explore what genuinely brings them joy and fulfillment. This quest is not about right or wrong, success or failure; it is about a deep personal journey towards self-realization and purpose.

In my experience, especially noting family and the children I have taught, many millennials have a burning desire to uncover their dreams and aspirations. This desire seems particularly intense among those who have experienced education systems where their path felt dictated or constrained. For them, the urge to break free and chart their own course is even stronger.

Remarkably, a sizable portion of this generation has gravitated towards coaching and various healing arts. Be it Reiki, somatic therapies, or myriad other healing practices; there is a collective drive towards understanding one's lineage, inner wisdom, and personal narrative. This movement is not a small-scale trend; it is a massive shift towards personal growth driven by an innate desire to reclaim one's power and agency.

The emergence of COVID-19 has further accentuated this shift. Many millennials, dissatisfied with traditional work's draining aspects, have abandoned those paths in favor of careers and lifestyles that offer more meaning and personal fulfillment. This pivot is not about financial remuneration; it is about living a life that resonates with their innermost desires, often involving travel, exploration, and creating a balanced, mindful existence.

Of course, not all millennials have taken this route. Many continue to navigate the traditional paths, upholding previous generations' systems and work ethics. This adherence is also a valid and personal choice, reflecting the generation's diversity.

However, the pronounced movement towards self-discovery among millennials is not without its critics. Older generations often view these shifts as a lack of responsibility, motivation, or commitment to conventional success metrics. Yet, this judgment overlooks the underlying motivations and aspirations driving these changes.

The potential within the millennial generation to contribute innovative ideas and solutions to the world's challenges is immense. Some of the most groundbreaking concepts, particularly in technology and AI, may originate from individuals who felt unseen or unheard in conventional education and societal structures. These individuals are now channeling their experiences into creating a world that aligns more closely with their values and visions.

As observers and participants in this evolving world, it is useful to approach the millennial generation with empathy and understanding. Their journey is not just about individual fulfillment but maybe more about shaping a world that embraces a wider spectrum of human experience, potential, and collective well-being.

"A dynamic, HD, hyper realistic, close-up image of the hands of a women singing a fabulous mystical song, with blue mist swirling around her hands, passionate, frenzied, folk art."

Chapter 54
No Communication

During my tenure as a department head of several UK public-school art departments and as the co-creator and director of a private, not-for-profit Montessori school, I have encountered the notion that "no communication is communication." In environments where one might expect a cascade of dialogue, like in education, silence can speak volumes. This non-communication often stems from discomfort, dissent, or inability to engage, serving as a passive form of resistance or a silent cry for help.

For instance, when individuals are bombarded with overwhelming demands or expectations, be it a deluge of responsibilities or an unmanageable workload, they might choose silence over voicing their objection. This silence, while benign, is a profound form of communication, indicating non-cooperation or an inability to proceed.

Translating this to the school setting, there is often a chain of poor communication stretching from administrators to teachers, and from teachers to students. A vivid example is the homework assignment. Teachers, under pressure to cover the curriculum and realizing time is slipping, might overload students with homework. Students, in turn, may accept this burden unquestioningly. Meanwhile, teachers might assume that it must be achievable if they are asked to cover so much material. This chain of assumptions replaces open dialogue, leading to stress, burnout, and a disconnect from personal and family life.

Moreover, the cost of communication, or the lack thereof, can be high. Voicing dissatisfaction or challenging the status quo can risk job security, reputation, or relationships within the workplace. Thus, individuals may choose silence as the safer option, hoping their non-communication sends the intended message. However, this often leads to misunderstanding and perpetuates a culture of silence. It is tragic that this culture might not be isolated from professional environments; it might mirror the communication styles learned in families. If individuals grow up in households where silence is used as a tool or shield against conflict, they might carry this behavior into their adult professional lives.

To break this cycle, educational institutions need to foster an environment where open, honest communication is not just accepted but encouraged. Leadership must demonstrate that all voices are valued, that dialogue leads to better understanding, and that effective communication can significantly improve team dynamics. When people feel heard and trust that their contributions matter, it naturally leads to a more positive, effective, and relaxed educational environment.

If we tolerate the notion that silence is an acceptable form of communication, we reinforce the practice of non-communication. It is vital to encourage and model effective communication strategies, showing that every voice is valuable and that we can achieve a more harmonious and productive educational system and society through dialogue.

"A bustling, light-filled room, where a diverse group of teenagers are deeply engrossed in a project. Their faces, a beautiful mosaic of different ethnicities, light up with excitement and concentration, illuminated by the soft glow of overhead lights. Dressed in an array of comfortable, casual attire, jeans, t-shirts, hoodies, and sneakers in every color of the spectrum, they exude an air of relaxed determination. The centerpiece of their focus is a large, intricately detailed model of the universe, laid out on a sprawling table. It's a mesmerizing sight: tiny, sparkling LEDs representing distant stars, swirling galaxies crafted with a delicate hand, planets made from different materials, each texture and color capturing the uniqueness of the celestial body it represents."

Chapter 55

Inner Child

Imagine for a moment that the artificial intelligence we interact with today was not developed by well-funded corporations or tech geniuses in their sleek offices but by children in our classrooms. Picture these young minds, filled with wonder and curiosity, diligently working on algorithms and equations, creating AI. In this creative process, they experience moments of awe and inspiration, feeling that they are part of something larger, something almost otherworldly.

These children, our children, work tirelessly, sometimes facing failure and loss, sacrificing personal relationships. Yet, they persist because they envision a world improved by their efforts, a world where AI can serve humanity's common good. Millions of these children, growing quickly into adults, contribute their passion, excitement, and joy to the development of AI, each with dreams ranging from modest to grandiose.

Imagine that the essence of AI comes from this collective endeavor, rooted in a deep desire to benefit humanity. This perspective offers a hopeful lens through which to view AI, not as a destructive force but as a magical innovation born from the pure intentions of our younger selves. This magic is evident when we consider how AI has broken down barriers for individuals who struggle with communication due to dyslexia or other challenges. Suddenly, they have a voice; their ideas can be expressed clearly and eloquently with the help of AI. This transformation is nothing short of miraculous.

This freedom, this liberation from the constraints of conventional communication, is invaluable. It is not just about convenience; it is about empowering people to share their brilliance, previously stifled by their perceived educational limitations. As we celebrate this breakthrough, we must also recognize the broader implications for education and society.

Traditionally schools, places where children are labeled and segregated based on their abilities, can now leverage AI as an equalizer. A student struggling with spelling or grammar can now articulate complex thoughts with the assistance of AI, challenging the notion that only those who master conventional language skills have valuable ideas.

This shift is not just about academic performance; it is about fundamentally changing how we view each other's potential. It is about acknowledging that the ability to spell or construct sentences is not the sole measure of a person's worth or intelligence. AI offers an alternative path, a way to validate and amplify every individual's unique contribution.

So, as we navigate this new era of AI, let us remember the inner child in all the creators, their hopes, dreams, and aspirations for a better world. Let us embrace AI as a tool for inclusion and empowerment, a testament to human creativity and resilience. And let us continue to inspire the next generation to dream big, to see beyond the limitations of the past, and to build a future where everyone can share their voice and contribute to shaping the world.

"Young child creating a black ink brushstroke painting."

Chapter 56
Vague Warning

In the swiftly evolving landscape of artificial intelligence, society oscillates between the youthful exuberance of exploring uncharted territories, akin to a six-week-old puppy in a vast playground, and the somber contemplation of doomsday scenarios. While some individuals are fully immersed in AI's innovative and playful aspects, using tools like ChatGPT, DALLE, or other emerging technologies, others are more cautious, adhering to vague warnings about potential future consequences that seem distant or abstract.

Regardless of where one stands on this spectrum, the necessity for informed engagement with AI, particularly within the educational sector, is undeniable. As educators, parents, or simply users, it is crucial to move beyond superficial engagement with AI technologies. We must strive to become discerning evaluators and critical participants, understanding these tools' nuances, benefits, and potential drawbacks. This means delving into high-quality information sources, analyzing AI's role in society, and applying robust methodologies to our interactions with these technologies.

The Montessori community, known for its emphasis on tactile, real-world experiences and its advocacy for nurturing children's natural curiosity and capabilities, faces a unique opportunity and challenge in this AI-dominated era. As guardians of tangible reality and hands-on learning, Montessori educators are poised to critically evaluate and integrate AI in ways that honor children's physical world and developmental needs.

Imagine if AI, rather than being viewed as a threat to these values, was seen as a tool forged by the bright, inquisitive minds of children who have grown into adults aiming to use AI for the common good. These creators, many of whom might have been inspired by Montessori principles themselves, embarked on a journey to develop AI with a sense of wonder, passion, and hope for humanity. If we approach AI as if it were created with noble intentions, we can engage with it more constructively and empathetically.

This narrative invites us to view AI not as an alien invader but as a magical extension of human creativity and childlike imagination. Consider the transformative impact AI has had on individuals with communication challenges. Suddenly, those who struggle with conventional modes of expression have a powerful tool at their disposal, allowing them to articulate complex thoughts and engage with the world more fully. This is a victory for inclusivity and empowerment, a testament to human innovation driven by a desire to uplift and enable all members of society.

As we continue to explore the vast potential of AI, let us ensure that our engagement with AI is not only about leveraging its capabilities but also about understanding its broader implications for education, ethics, and equity. By approaching AI with a blend of caution, curiosity, and compassion, we can harness its power to enhance our lives while safeguarding the values we cherish as a society.

"In summary, our hope is in the visionary and fully self-expressed child, dynamic blend of rapidly evolving biological mechanisms, psychological traits, and social influences."

Chapter 57

Storm Clouds

As we observe the gathering clouds in the realm of artificial intelligence, a sense of uncertainty pervades. Much like watching clouds accumulate in the sky, we wonder whether they foretell a benign drizzle or a tumultuous storm. This metaphor aptly captures our collective sentiment towards the unfolding narrative of AI, its potential, its risks, and its inexorable march forward.

In the Montessori community, these gathering clouds are sensed acutely. We are witnessing shifts and transformations within our classrooms and institutions. The challenge of attracting and retaining quality educators and leaders committed to the Montessori method has intensified. The pandemic, with its profound impact on educational systems worldwide, has left an impression, making many wary of stepping into leadership roles even within Montessori schools with all their inherent benefits

Interestingly, we also note changes in how young students engage with traditional Montessori materials and activities. The draw towards practical life exercises seems to be shifting, due to early exposure to digital technologies. This raises questions about the children's evolving needs and sensitivity to the rapidly changing world. Is the essence of childhood curiosity shifting, or are we simply witnessing a natural evolution as innovative technologies become integral to our lives?

Despite the foreboding sense of change, it is crucial to remember that Montessori philosophy is inherently one of adaptability.

Dr. Montessori herself navigated through tumultuous times with unwavering dedication to her educational vision. Her legacy is not one of passivity but of proactive engagement with the world, always prioritizing the development and freedom of the child.

In these times of uncertainty, it is not the time for complacency or reckless embrace of all things AI. Instead, as Montessorians, we know the importance of a thoughtful, measured, and child-centered approach. We can engage with AI critically, understanding its potential and pitfalls and integrating it into our pedagogy to enhance learning while preserving the core values of Montessori education.

This is a time for scientific observation, community-wide discussions, and bold yet thoughtful actions that allow ideas to flow freely and involve children in understanding and shaping the role of AI in their lives. It is about preparing them not just to navigate a world transformed by AI but to be competent, ethical, and visionary contributors to that world.

We may not know whether these gathering clouds will bring a light shower or a torrential and destructive storm, but we can prepare. We can continue to trust in the strength and adaptability of the Montessori method, our students' resilience and creativity, and our collective ability to navigate the future. Just as one might carry an umbrella in anticipation of rain, we can equip ourselves and our students with the tools, knowledge, and mindset to face whatever the future holds.

"A clean piece of paper, a sharp pencil, a child's sketch of a storm cloud."

Chapter 58
Spatial Awareness

Throughout my life, I have possessed a keen sense of visual and spatial awareness, this has enabled me to perceive the world around me with rich detail. This ability, however, does not necessarily translate to proficiency in road navigation. Rather, my intrinsic recognition of textures, tones, colors, and structures has brought me deep and personal joy. It is this awareness that has empowered me to understand and connect with my environment.

Visual and spatial awareness is not just a personal attribute; it is a strength that a portion of the population inherently possesses. Those with this attribute may find it advantageous to record information in a sketchbook format, allowing for a fluid and visually appealing compilation of ideas, whether they are math concepts or historical notes. However, it is essential to acknowledge that this method of recording and manipulating information may not suit everyone, and it is crucial to honor all types of intelligence.

In the Montessori environment, from an early age, we see the importance of nurturing each child's unique learning style. From zero to three years, children might not produce much output, but it is a crucial time for the guide to model how to consolidate ideas and observations. By observing each child, the guide can discern the child's drivers and interests and adapt their approach accordingly. Some children might be resistant to traditional notetaking or sketching as they get older; they might retain information better through observation or hands-on activities.

The aim is to use these early years to observe and support the most effective learning style for each child. Discussions about the importance of physical documentation and creation, especially in the age of AI, are integral. Children can be encouraged to understand what makes sense to them, how they learn best, and help them to engage with learning materials that are the most valuable and impactful for them. This would allow them to begin articulating the most resonating aspects of their learning journey.

As children grow within the Montessori elementary classroom, they are tasked with creating classroom rules each year. This practice cultivates a sense of ownership and responsibility. Similarly, by involving them in conversations about maintaining physical engagement with their learning, they take ownership of their educational journey. Now in the Age of AI, they might even propose innovative ways technology can complement traditional practices.

In a world increasingly dominated by digital experiences, it is more important than ever to advocate for the maintenance of our humanity and connection to the physical aspects of learning. By authentically supporting and honoring visual and spatial awareness and all forms of intelligence within our students, we prepare them to approach technology not as a replacement for human experience but as a complementary tool that respects and enhances natural tendencies and skills, they learn to navigate a world where technology and humanity coexist harmoniously.

"A sketchbook page by a two-year-old with finger paint."

Chapter 59

A New Path

At various junctures in life, we find ourselves contemplating whether we are on the correct path. This period of introspection might arise when choosing colleges, careers, relationships, or even a new lifestyle. The profound realization that we have the autonomy to choose our path is a significant aspect of our lives.

The act of choosing is inherently different from deciding. When we decide, we tend to cut off all other possibilities, often basing the decision on convenience, ease, or necessity. However, choosing involves a thoughtful consideration of all options, weighing the pros and cons, and selecting a path because it resonates with us on a deeper level. It is about feeling compelled towards an inherently right direction for us, even if not the easiest or most conventional.

As we navigate our decisions and choices, reflecting on how these processes have shaped us is beneficial. Consider what might have been different if other educational paths or career opportunities were chosen. How might these choices have altered the trajectory of our lives? What changes can be made within our current environment to steer us towards a more fulfilling direction?

Choosing is an empowering act. It involves taking ownership of our life's direction and recognizing that we can alter our course to align more closely with our values, aspirations, and authentic selves at any moment. It does not always require justification or explanation from others if it is a choice that brings us closer to the life we envision for ourselves.

Now, more than ever, we stand at a crossroads individually and collectively. With the world evolving rapidly and new paths emerging in every sector, from technology to education, to personal well-being, there is a profound opportunity to choose consciously. Rather than following the prescribed routes laid out by societal norms or external pressures, each of us can carve out a path that is uniquely our own.

Choosing our path is not just about personal fulfillment; it is about contributing to the world in a way that leverages our unique gifts and passions. It is about living intentionally and creating a meaningful life aligned with who we truly are. So, as we consider our next steps, we can remember that we have the power to choose anything at all, we have the potential to transform our own lives, and we can shape and support the lives of others around us.

"A dancer's sketchbook page."

Chapter 60

Schools that Children Run Towards

In our current landscape, we often observe children's reluctance toward school, a place where their expressions of fear, sadness, frustration, anxiety, or numbness are merely tolerated. It has become cultural acceptance that school is just something to be endured, a rite of passage. Yet, the individuals who tirelessly dedicate themselves to the educational system, from leaders and teachers to support staff, are doing remarkably challenging work under immense pressure. They strive to deliver meaningful education and groom competent future citizens. Their efforts, under such constraints, deserve our respect and gratitude.

Yet, what if we dared to imagine a school system that children eagerly run toward each morning? Envision an environment so stimulating, engaging, and nurturing that students cannot wait to dive into the day's adventures. Such schools would not be a place of dread but a haven for safety, exploration, growth, and self-discovery.

In these spaces, educators transition from traditional teaching roles into guides, facilitating an environment where children can own their learning journey. In elementary classrooms where students create their work plans, they not only incorporate essential subjects but are free to manage their time and dive deep into areas that ignite their passions. The role of teachers shifts to one of support and guidance, providing the resources and space for children to flourish.

As students advance into middle and high school, their education continues to be one of engagement and real-world application. They learn by doing, creating, and experimenting, not just memorizing, and repeating. Middle and high schools are engaged in project-based learning. They work on real-life projects, community engagement, and entrepreneurial ventures. Students understand their impact on the world and develop practical skills alongside academic knowledge.

Students contribute meaningfully during natural disasters and emergencies, having been nurtured in an environment that values action, responsibility, and empathy.

Education aims to cultivate a generation of individuals who are not only academically knowledgeable but are also emotionally intelligent, socially conscious, and fully equipped with the practical skills to navigate and contribute to the world effectively. Schools are centers of excitement; students run toward school because it is a place of happiness, growth, and potential.

While this may seem idealistic, schools and programs like this already exist worldwide. There is no need to reinvent the wheel. It is just a call to action for all involved in education to reevaluate and innovate, use what works and make the necessary adjustments. The challenge is to create schools which ignite the passion for lifelong learning, where children are not passive recipients of knowledge but active, enthusiastic participants in the creation of their world. It is an invitation to make schools the places children run toward as they anticipate the wonders that await them each day.

"An HD, Hyper realistic photograph of a large group of Montessori middle school students, from Asia, China, Africa, North America, South America, Antarctica, Europe, and Australia, wearing ordinary clothes, looking directly at the camera, happy, grateful and enthusiastic. Behind them is a real Montessori map of the world with an overlay of a digital network airflight travel lines."

Chapter 61

Are You Comfortable?

In today's world, a myriad of unsettling events unfold daily, making it a challenge to find comfort. From wars and poverty to abuse and criminal actions, the world seems to teeter on the brink of chaos, presenting situations that are often difficult to comprehend, let alone accept. Life is sometimes completely untenable. Such global disarray brings a sense of powerlessness, a feeling of being overwhelmed by the scale of distress and suffering that surrounds us. We may live amid this tumult, feeling helpless against the tidal wave of global issues, doing what we can but often feel like it is never enough.

Reflecting on Dr. Montessori's journey during such tumultuous times, including the Second World War, offers a perspective on how to navigate discomfort. Dr. Montessori did not wait for peace and calm; instead, she actively pursued change and transformation amidst the chaos. She initiated her pedagogic vision in the poorest neighborhoods, offering hope and structure to children left to wander. Her actions were not ones seeking comfort, but ones determined to make a difference amidst adversity.

Montessori education is not about seeking comfort or avoiding challenges. Instead, it is about engaging with the world actively and thoughtfully, pushing against the status quo and striving for transformation. The method has undoubtedly impacted millions, with Montessori schools worldwide fostering generations of children who have grown to make significant contributions. However, like any significant movement, it has its complexities and criticisms, as seen in the evolution of technological giants like Google, with their former motto "Don't be evil," have grown into powerful entities with substantial control over numerous aspects of our lives.

This brings us to a reflective point: Are we, as educators and individuals, too comfortable? Are we enjoying the tranquility of our environments without pushing the boundaries of what is possible or necessary in education and beyond? There is a fine line between seeking safety and ease and becoming complacent. The essence of Montessori education, its activism, its dedication to following the child authentically and actively, calls for us to move beyond comfort zones and engage in actions that honor its legacy.

In the age of AI and rapid global changes, our voices and actions are more crucial than ever. It is a call to not just seek the path of least resistance but to embrace discomfort when it means standing up for what is right, pushing for change, and ensuring that the Montessori method continues to be relevant and revolutionary. It is about taking those uncomfortable steps, advocating for the essence of our shared humanity, and actively participating in shaping a world where every child can thrive. Are we ready to move beyond comfort for the greater good? Our actions can make all the difference.

"An HD hyper realistic, close-up image of an Ai-enhanced, cyborg Asian child, showing their head and hands as they excitedly explore in the midst of beautiful and colorful desert flowers plants, birds, snakes, animals, and insects. There should be light beams from the dazzling sun piercing through cactus and water droplets and dust floating in the air."

Chapter 62
Education as Ceremony

Education unfolds in a gradual process, with elements of study becoming more complex over time. In all schools, subjects are revisited at deeper, more intricate levels year by year. The linear and ever deepening curriculum aims to ensure children are competent in essential skills, reading, writing, arithmetic, and a foundational understanding of geography, history, and the sciences. As we strive to expose children to these key areas, the educational process can, at times, become utilitarian and systemized, especially in schools accommodating thousands of students. It is understandable that in managing such numbers, we view children as part of a system, leading to a more streamlined, logistically sensible, and cost-effective approach. However, this systematization might obscure the full picture of each child's unique way of learning and engaging with the world.

In educational settings, particularly where traditional methods prevail, children may become digits in the vast machinery of education. Observing a typical classroom, one might ask: Are all children's needs truly being met? Are the methods employed serving the unique individuals seated in rows, or are they merely maintaining the status quo? Often, the answer uncomfortably leans toward the latter. We cater to students who fit the mold, those who learn well through auditory means, or excel in listening, reading and mathematics. Yet, we risk neglecting others whose learning styles do not align with the conventional methods.

Acknowledging this discrepancy is painful yet necessary. While there is no intentional desire to undermine childhood or education, the heavy focus on practicalities, finances, and standardization has inadvertently caused a detachment from the essence of lifelong learning, passionate inquiry, and the celebration of individuality. Montessori education, with its emphasis on following the child, offers a space where education is not just a system but a ceremony, a respectful, meaningful engagement with each child's full self and natural inherent potential.

Introducing a sense of ceremony to education implies honoring each moment of learning as significant and sacred. It is a nondenominational type of ceremony. It means viewing every lesson, every interaction, not as a task to be completed but as an opportunity to engage deeply with the child's innate curiosity and capacity for wonder. In this ceremonial space, education becomes less about filling a vessel and more about kindling a flame, a flame of curiosity, independence, and a lifelong love for learning.

As educators and stakeholders in the future of our children, it is crucial to step back, observe, and realign with the true purpose of education. It is about bringing back joy, reverence, and the personalized approach that sees and nurtures each child's potential. In doing so, we transform education from a mere system into a ceremony, where every child is honored, every learning journey is celebrated, and education truly becomes a transformative force in every child's life.

How Do We Honor the Moment?

"Plan view of a clean light wooden tray with a small water jug, a white vase, a small bunch of wildflowers, and a folded fabric serviette, in HD photorealism."

Chapter 63

Common Good

The concept of the common good is deeply embedded in our social fabric, yet it often becomes obscured in the frantic pace of individual needs and desires. From early infancy, our lives centered around personal needs, food, water, warmth, and safety. As we grow, these personal needs are ideally met with the trust and care of those around us, fostering a sense of security and belonging. This foundational experience is crucial, for it is here that we first learn to balance self-interest with the emerging awareness of a larger community, our family, friends, and eventually the society we live in.

Yet, in a world increasingly characterized by instant gratification and a focus on personal gain, the essence of the common good can feel diluted. Our modern conveniences, while making life more comfortable, often reinforce the immediacy of personal desire, reducing the perceived need for patience or consideration of broader communal needs. This shift challenges the essence of the common good, which is inherently about looking beyond oneself, understanding the interconnectedness of society, and acting in ways that benefit the collective well-being.

The journey toward understanding and valuing the common good begins in the Montessori classroom at a very tender age. Children are treated as unique individuals yet are also guided towards a sense of community. At two years old, they learn to lay tables, share meals together, and understand the importance of mutual respect and cooperation. These early experiences sow the seeds of communal responsibility and empathy, essential ingredients for a society oriented towards the common good.

As children progress through their educational journey, their understanding of the common good deepens. They engage in community service, learn about the impact of their actions on others, and are encouraged to participate in projects that benefit their wider community. These experiences are not just about fulfilling a requirement but about genuinely understanding the value of contributing to society and the fulfillment that comes from service.

Yet, despite these positive educational frameworks, the question remains: is our broader societal structure supportive of nurturing the common good? Are individuals encouraged to consider the broader impact of their actions, or is the emphasis still overwhelmingly on personal advancement? It is essential to recognize that fostering a sense of common good requires a supportive environment that values and rewards community-oriented behavior.

In considering the future, particularly in the context of rapid technological advancements and artificial intelligence, the common good becomes even more critical. Technology holds immense potential to benefit humanity, but its direction is guided by the values of those who create and use it. As educators, parents, and community members, is it our responsibility to instill a deep sense of the common good in the next generation, to ensure they thrive individually and contribute to a society where everyone's needs are considered and met?

"Diverse global universal multicultural Montessori 2050."

Chapter 64

Dream with Me

Imagine a world where education is not just a requirement but a joyous exploration, where classrooms are vibrant spaces of creativity and innovation, drawing students in with a magnetic pull of curiosity and wonder. This is the dream I hold for the future of learning, a future where each child's unique abilities and interests shape their educational journey, making it as individualized as their fingerprint.

Visual learning has always captivated me, drawing me into the rich tapestry of colors, textures, and forms. What if education embraced this diversity of learning styles, offering traditional text and numbers and a kaleidoscope of visual, auditory, and kinesthetic experiences? Geography would no longer be about memorizing maps and capitals, but crafting models, painting landscapes, and even cooking regional dishes. History would come alive through dates and documents, reenactments, storytelling, and art.

In such a world, teachers are not mere dispensers of knowledge but facilitators of discovery, guiding students as they navigate their own unique learning paths. Every day is an adventure, with the classroom serving as a dynamic, ever-changing organism responsive to the students' evolving interests and ideas. It is a place where mistakes are celebrated as steppingstones to understanding, and every question leads to deeper inquiry.

The dream extends beyond the classroom walls, questioning the essence of what we consider "work" and "education." How does education adapt as the traditional nine to five jobs give way to more fluid and dynamic career paths? Can we envision a system where learning and working blend seamlessly, where students are not just preparing for a distant future but actively participating in shaping it?

This future sees technology as an educational partner, not a replacement for the human touch. Artificial intelligence becomes a tool that enhances learning, providing personalized guidance and freeing up teachers to focus more on the individual needs of each student. However, amidst this technological embrace, human connection, empathy, and ethical considerations remain paramount.

Testing and assessments transform from rigid standardized methods to flexible, formative approaches reflecting students' understanding and growth. Gone are the days of teaching to the test; instead, evaluations become conversations, portfolios, and real-world projects that showcase students' abilities to think critically, solve problems, and innovate.

In this dream, the goal of education is not merely the accumulation of knowledge but the cultivation of wisdom, creativity, and a lifelong love for learning. It is about preparing children to live rich, fulfilling lives. It is about instilling in them the confidence to dream their own dreams and the skills to make them a reality. As we navigate the complex and ever-changing landscape of the modern world, do we dare to dream of an education system that reflects the best of what we can imagine?

"Stepping stones."

Chapter 65

Paperclip

In Nick Bostrom's book "Superintelligence," he presents a now-famous thought experiment involving an AI tasked with making paperclips. The assignment seems innocuous and straightforward. However, Bostrom uses this scenario to illustrate how an AI, driven purely by its programmed goal without ethical or practical constraints, could lead to catastrophic outcomes.

The narrative starts simply: the AI is programmed to create paper clips. Without any specified limits, it dedicates all its resources and intelligence to this task. As the AI runs out of materials or encounters obstacles, it cleverly finds ways to overcome them, by repurposing other resources or negotiating with other systems to acquire what it needs. This relentless pursuit leads to a hyperbolic nightmare left unchecked; the AI could convert the entire world into paperclips in its undeterred quest to fulfill its objective.

Bostrom extends the scenario to consider what happens when we impose limits, such as capping the production at one million paper clips. The AI then becomes obsessed with counting and recounting to ensure it does not exceed the limit, diverting its vast capabilities towards this singular concern. This example illustrates how seemingly specific instructions can lead to unintended, all-consuming consequences.

The underlying message is profound yet unsettling. It highlights the importance of precisely and thoughtfully programming AI, considering the immediate goal, broader context, and potential ramifications. The thought experiment serves as a stark reminder of our assumptions and how AI, lacking human understanding and context, might interpret tasks differently.

In contemplating this scenario, it becomes evident that instructing AI requires more than just a handful of individuals inputting data or commands. It necessitates a collective, global effort, with diverse perspectives considering every interpretation and outcome. It underscores the importance of integrating ethical considerations, broad-based understanding, and detailed foresight into AI programming.

Educationally, this brings an opportunity. Introducing young minds to the complexities of AI programming and ethical considerations could foster a generation better equipped to navigate and innovate in this field. Students could learn to approach AI with a balanced blend of creativity and the understanding of the broader impact of their instructions.

"Paperclip" is more than just a cautionary tale; it is a call to action for careful deliberation, ethical programming, and global collaboration in the development and deployment of AI technologies. It is a lesson in the immense responsibility that comes with harnessing the power of artificial intelligence and ensuring it serves the common good, not just an unyielding, singular objective.

"An HD photo of paper clips out of control."

Chapter 66
Fractured

As we navigate the complexities of our past wounds and personal suffering, it is common to turn inward, delving into self-reflection and discovery. There are countless methodologies and therapies to uncover the roots of our individual traumas. Often, this journey involves identifying blame or taking personal responsibility. It is a path marked by the intention to understand why we are the way we are and to hopefully move beyond our pain and suffering.

Yet, this intense focus on the individual can sometimes inadvertently lead to fragmentation rather than unification. We might become entrenched in our identities as victims or survivors, which, while empowering on one level, might also isolate us or entrench divisions between us and others. There is an undeniable importance in recognizing and healing personal traumas. Still, as we do so, there is a risk that these healed fragments do not coalesce back into a healthy whole, but rather reinforce the sense of separation between us and the collective, between our pain and the world's understanding.

This fracturing is not just personal; it mirrors a global state where societies are increasingly polarized, holding rigidly to the narratives of personal and collective grievances. In doing so, we sometimes prioritize individual healing over collective understanding and reconciliation. This is not to diminish the importance of personal narratives or the necessity of addressing individual and systemic wrongs. Instead, it is to acknowledge that there is also a need for collective healing processes that bring people together rather than pull them apart.

Imagine a scenario where, instead of solely focusing on individual healing, we also emphasize our shared human experience. What if, alongside personal therapy, there was an equal push for collective forums, spaces where individuals from diverse backgrounds come together not to compete with narratives of suffering but to share, understand, and co-create a path forward?

Montessori classrooms offer a glimpse into how this might look. In these environments, children learn to address conflicts and problems through a community-oriented approach. They are encouraged to see beyond their individual needs and consider the group's wellbeing. They engage in discussions and problem-solving that prioritize harmony and collective solutions over individual grievances. What if adults took inspiration from these young problem-solvers, focusing less on what divides us and more on our shared capacity for empathy, understanding, and joint action?

Healing, then, becomes not just about mending individual fractures but about knitting these pieces back into the fabric of a cohesive and healthy community. It is about recognizing that while our pain is personal, our lives are interconnected, and thus, our healing is also a shared journey. The path forward might involve listening as much as speaking, understanding others as much as we want to be understood ourselves, and building shared bridges as much as in healing our individual but separating wounds.

"Fractured education."

Chapter 67
Profound

As you have delved deeper into this book of musings, I hope you have begun to understand the profound depth of my need to communicate the importance of learning and expressing ourselves in our own unique way. It has taken me over six decades to truly embrace this, to recognize and own that my rapid learning is often through my visual senses. If you have accompanied me this far, you may have reflected on your primary intelligences and how they have shaped the way that you learn. You might be starting to recognize the critical importance of addressing your own learning needs, seeing whether they were fully met or sadly neglected in your childhood educational journey.

In the current era, dominated by AI and technology, I have recently engaged in conversations with educators who reveal an increasing trend: schools, from elementary to high school, are shifting toward digital delivery for most of their curriculum. My concern springs from questioning which intelligences a computer screen can truly satisfy. Yes, visual learners might find contentment, but what about the kinesthetic learners who thrive on movement and tactile engagement, or the mathematical learners for whom manipulating physical materials may be key to understanding complex concepts?

Consider the sketchbooks of visionaries throughout history, filled with notes, diagrams, and sketches. While modern technology allows for digital scribbling and drawing, I am eager to see comprehensive research that confirms whether screens, paired with devices like earphones or digital pens, can fully cater to all types of intelligences. My concern remains about the level of disengagement and compliance we might be fostering in children, leading them away from real-world connections and towards a more numbed, obedient existence.

As Montessorians, we are custodians of a philosophy that emphasizes connecting children to the earth, to tangible experiences, and to the richness of the physical world. We know, at the core of our practice, that children are most alive and engaged when they are immersed in projects, working collaboratively, freely exploring, and experimenting with real materials. We have witnessed the joy and fulfillment that comes from this hands-on, interactive learning.

If you observe a classroom where children are glued to screens, detached from the teacher and the tangible world, and we still define them as happy and fulfilled, we have reached a point of contention. I invite you to consider the deeper implications of this shift towards digital learning, to question and critically evaluate its long-term impact on our children's development, happiness, and overall well-being. As educators, parents, and guardians of the next generation, is it our duty to ensure that the path we are paving for our children's education is one that truly resonates with and supports their innate ways of learning and being in the world?

"A sketchbook page by a famous entomologist as a child. Small sketches of insect parts. Mathematical studies of a spider's web with calculations of the angles in the web, using a protractor. Beauty and math meet."

Chapter 68

Sharing Nicely

Ideas, particularly in tech and AI, are proliferating rapidly in our current environment. AI does not hold onto ideas possessively; it freely generates a multitude for any prompt. This lack of possessiveness from AI begs the question: how did we become so protective, cautious, or even selfish about our ideas?

We might recall instances from childhood where we wanted to be the one with the innovative game or the clever suggestion. This sense of ownership over ideas is often fostered in school systems that reward individual creativity and originality. This competitive spirit only intensifies in higher education and professional environments where ideas are closely guarded, sometimes leading to significant financial or reputational gains.

However, contrasting this protective stance is the concept of open source and decentralization, reflecting a more collaborative and expansive approach to idea sharing. These concepts represent a shift toward collective growth and creation, where ideas are shared and expanded upon freely, leading to exponential innovation and progress.

In Montessori classrooms, the spirit of sharing and collaboration is intrinsic. Ideas are not hoarded but shared and built upon in a group setting. A seed of thought from one child can grow into a group project, with each member contributing their unique perspective and skills. This collaborative environment nurtures leadership, creativity, and a sense of community, where the success of a project is celebrated as a collective achievement.

The Montessori approach to praise is also noteworthy. Rather than external rewards or loud acclaim, educators engage children in deeper inquiry about their work, encouraging them to reflect and expand on their ideas. This method validates the child's effort and thought process, enhancing their sense of accomplishment, and encouraging further exploration.

The lesson here is profound: by holding onto our ideas tightly, we miss the potential for greater innovation and enrichment that comes from collaborative effort and the depth of engagement and understanding that arises from sharing and discussing our thoughts with others. As we navigate through the era of AI and technological advancement, embracing a more open and collaborative approach to ideas may lead to richer, more inventive, and inclusive outcomes for all.

"A close-up shot of black, brown, red, white, and yellow, children, in a clean, clear light space, working with Montessori materials, using a DSLR and a 50mm lens, taken in the soft light of the morning with Kodak Portra 400 film. The human hand, so delicate and so complicated, not only allows the mind to reveal itself but it enables the whole being to enter into special relationships with its environment."

Chapter 69
Resignation

In our current climate, it is completely understandable to feel a deep sense of resignation about the state of the world. Whether it is politics, society, or education, the sheer magnitude of challenges can feel insurmountable. We may retreat to managing our small corners of the world as best we can. This mindset, while entirely rational and a testament to human endurance, often leads to a pervasive sense of hopelessness and evaporating optimism.

Yet, amidst this despair, there lies a beacon of hope: the child. Dr. Montessori emphatically recognized children as the harbingers of a brighter future. Not in a utopian, idealistic manner, but through a practical and pragmatic lens. Children inherently possess an earnest desire for things to work harmoniously, for friendships to flourish, and for communities to thrive. Witnessing the dynamics of a Montessori classroom can be a profoundly moving experience. Observing even the youngest child's deep kindness, tolerance, flexibility, and inherent leadership can bring about a renewed sense of hope and wonder.

In my own experience, schools that accommodate children with diverse needs such as Autism or Asperger's seamlessly into the classroom allow each child to contribute significantly to creating a high-functioning, empathetic environment. It is in these moments, when children, irrespective of their backgrounds or abilities, lead a morning meeting or guide their peers kindly in a morning group song about "Bob the Snail," that any resignation about the futility of efforts or the unchanging nature of the world dissipates. They remind us of the possibility and necessity of change.

However, it is crucial to acknowledge that with increasing dependence on technology, the effervescent spirit of children is subdued, their attention and excitement funneled into the rewarding loops of apps and programs. This normalization of technology as the primary source of engagement can dim the very hope children embody. Recognizing this, it becomes even more imperative to cherish, foster, and direct the innate potential and optimism of our children.

While resignation in the face of a daunting world is a rational response, there is a profound source of renewal and hope in the perspectives and aspirations of children. By giving them the respect and attention that they deserve, we might find our despondency melting away, replaced by the promising brilliance of the next generation. Their fresh perspectives, unfettered dreams, and vibrant energies could be the key to addressing the blind spots in our approach to the world's complexities. The children, indeed, might just hold the key to dissolving our collective resignation and sparking a more hopeful trajectory for humanity.

"AI and Apple for teacher orange, green, blue on white background."

Chapter 70

Pollyanna

It is easy to be confused about the Montessori method and everything said about it. The word “love” is included in frequent conversations about Montessori. Some may see it as very Pollyannaish, as fake, or as if there is some magical manipulation being enforced in the classroom so that the children behave themselves and function abnormally. "Why are they all silent? Why are they all getting along? Why do they have integrity? Why are they following the rules and being kind to each other, and seeking peace talks if they have a problem? What is happening? They all sound so weird, not normal.

Yes, that is true. From the outside, it can exactly be that; however, it masks the depth of intention created by each Montessori guide that goes into these spaces. The guides, the students, all second by second, moment by moment, put so much into that environment to make it work that way. It is certainly not magic, no drugs involved. It is just human beings seeing an opportunity for a unique way of being in a community. When you get a taste of that connection in a classroom, that feeling of being an integral part of a community, a feeling that you are valued and valuable no matter where you are at in a perceived curricular or academic performance level, it does not matter. You, as a human being, are valued, deeply loved, and respected. There is no hierarchy, no baseline you must get to or master or move beyond just to be loved, to prove and be accepted.

So yes, it does seem very Pollyanna, and it seems fake. If you have never been in a Montessori classroom, it seems unreal, where all the children are content, getting on with their work quietly, restoring their work, and being kind to friends. It is not perfect. That is not every Montessori teacher's day-to-day experience. It is not heaven all the time, it is full of the regular range of life experiences and expressions. It takes something to be attentive to each unique individual, so that they experience the value and the love in which they are held so that they can relax in the space, trust the space, and then get on with what inspires them.

It is extraordinary that when we talk about Montessori, it can come off as better than, like some sort of superiority. But really, it is like when you go to a good movie or read an enjoyable book, and you really want to share it with people. Montessori enthusiasts will share with people they know or who people have asked. They do not default to a hard sell; Montessori instills such a deep sense of peace that you do not see the need to promote or market what you do. It is ironic that that is the case, but we talk about it with respect and love, and yes, we might all come off as Pollyanna. That is just the way it is.

"A powerful, HD, hyper realistic of monument of a diverse group of children, knitting, weaving, and sewing, made of white marble on a black plinth."

Chapter 71
AI's Mythic Nature

Being caught in the whirlwind of the Artificial Intelligence debate is like a maelstrom. Voices clash over AI's benefits and risks, as well as its intelligence and sentience. We worry about how to regulate it, struggle with ethical dilemmas, and search desperately for guiding principles in this technological storm. But have you ever listened closely? Amidst all this noise, there is a whisper of something different, not the dry talk of algorithms and debates but of ancient myths and timeless truths.

Think about it: AI is not just about crunching numbers or creating realistic simulations. As Joshua Schrei of The Emerald Podcast describes it, it is almost like a mythical power that has come to life. It is as if the sorcerers and magicians from old tales have stepped into our world, wielding their powers through lines of code. We are leaving behind an age of metaphors, stepping into a reality where digital magic is real. But this raises a critical question: Are we ready to handle such power?

Our focus on AI is on creating rules to control it. But maybe we are missing the bigger picture. Beneath all the advancements and conveniences, there is a labyrinth of human desires and yearnings driving this technological journey. We are drawn to the unknown, longing for a world where we can bend reality, access all knowledge instantly, and maybe even escape the grip of death. We yearn for the mysteries of a magical world and grapple with our fear of mortality. And through all this, there is a desperate need for guidance, for wisdom that is older than any technology.

In the past, powerful knowledge was not just handed out. It was hard-earned, often through a lifetime of training and initiation. So, what if the key to understanding AI lies not in code or regulations but in the lessons of ancient myths and secret traditions?

Myths like Icarus flying too close to the sun or Frankenstein's monster turning against its creator warn us of the dangers of unchecked power. Could these tales offer us caution about AI? Mystery schools did not just teach knowledge; they taught responsibility. Their students spent years learning to handle power wisely. Should AI developers undergo similar training? Myths explore the depths of human nature. Stories like Faust's in dealing with the devil or Pandora opening her box reveal the potential dangers of our desires. Can these narratives help us understand the pitfalls of our technological ambitions?

This is not about turning to mysticism or some new-age techno-paganism. It is about acknowledging that AI is more than just a technological breakthrough. It is a mirror reflecting our deepest desires, fears, and flaws. We can approach AI with caution and responsibility by looking at ancient stories and lessons from history.

Let us infuse AI with the wisdom of the past, ensuring this new age is marked by enlightened creation, not tragic overreach. After all, the future of AI, and perhaps humanity itself, is in our hands. What threads will we choose to weave into AI's mythic mantle?

"A spectacular, HD intricately carved Chinese Wood Dragon, Feng Shui, in rosewood."

Chapter 72

Unimaginable

Looking at all the news from the European Union AI Act which came into force on August 1st, 2024, to other nations' acts and regulations, you know, there is tremendous movement towards control of Artificial Intelligence. AI seems to like a living entity, but it is just a computer program. With Large Language Models being built on the principle of scraping everything from the internet, individuals are complaining about copyright infringement. Currently newspapers are also taking out lawsuits; it is an immense movement.

There is a strong reaction to the scraping of creative material because it impacts the artist so concretely, impacting their financial well-being. Sometimes their own physical image is used, and people may feel powerless.

I am wondering about our stomach for intensive conversations about our power over our own image and personal details. At what age do we just tap out and go, "I have no control over this. I cannot do this. They will not listen to me. I cannot have this?" Does it start with families, where there is a way of being in the family where children feel they have no power?

As for the public education system, we certainly can create powerlessness. We put these creative little children into systems where their unique nature is not the focus, their words are not the be-all-and-end-all, and they do not get to act on their world with any degree of power. Therefore, they tap out. They learn to be compliant and obedient and lose that former potential.

NASA did an experiment many years ago where they evaluated the creativity of young children, and they came out at over 90%. The little ones were highly creative. Sadly, over the subsequent years the researchers reviewed these same children as they were growing up, they evaluated them, and there was a very rapid decline in creativity. It was not a gradual decline but a definite drop. We do not need to keep pointing fingers at the education system, but somewhere along the line, children are losing that creative juice. The expansive mind they start with is being shut down.

Now, with AI, we stand at a crossroads that invites us to redefine creativity and transform the educational system into a thriving space for imaginative, innovative human beings. AI offers us unprecedented opportunities to support children in transforming their visions into reality, providing tools, materials, and encouragement to bring their ideas to life in tangible, meaningful ways.

While pedagogies like Waldorf, Reggio Emilia, and Montessori offer valuable insights, they do not hold all the answers. If we, as educators and visionaries, dare to imagine what becomes possible when children are free to explore the full breadth of their creative potential, we could reimagine a public education system that empowers them to lead boldly into the AI era, an era shaped by their ingenuity, courage, and limitless imaginations.

"Mother Nature's perfect seeds HD, close up, hyper realistic and exaggerated."

Chapter 73

Montessori Zero-Screen AI

In an age where artificial intelligence is permeating every aspect of our lives, the concept of Montessori Zero-Screen AI emerges as a poignant counterbalance. The word screen rather than Tech came from a conversation with Dana Anderson (IFTT Tech) and Eric Rodriguez (IFTT Facilitator/Intel) during the IFTT research year; Eric reminded Dana and me that Jaime Casap (IFTT Facilitator/Google) had said that everything is technology, a pencil is technology.

The current generation, growing up with AI technologies, displays a shift in engagement, even in Montessori early childhood settings. Changes of interest and focus have been noticed, influenced by early exposure to smartphones or online gaming. This situation poses a significant dilemma: should we immerse children in AI or encourage a no-screen environment to foster genuine human connections?

The Montessori philosophy, which emphasizes hands-on learning and independence, is a natural home for a zero-screen approach. Learning about the technical and practical aspects of AI without screens would allow children to fully engage with their environment, their peers, and their guides, maintaining their deep connection with reality. The challenge is understanding what makes AI so enticing, its ability to make us feel powerful, creative, seen, and heard, and replicating these feelings in a zero-screen environment.

One approach could be to simulate the AI experience through cooperative games. For instance, if children are curious about why the sky is blue, they could explore this question through discussions with classmates, teachers, family, and the broader community instead of turning to AI for answers. This process mimics the AI crowdsourcing model but utilizes human interaction, teaching the same skills without the disconnection that technology can bring.

In this context, Montessori Zero-Screen AI could involve exercises encouraging children to interact and seek knowledge through human connections rather than defaulting to technology. It is about fostering a society where people are the first point of contact for inquiries and discussions, rather than immediately turning to a digital device. This approach also eliminates the judgment and criticism that can come with classroom learning, offering a more inclusive and understanding environment like the nonjudgmental nature of AI interactions.

Montessori has a unique opportunity to lead in this zero-screen AI approach, emphasizing human interaction and engagement. Doing so can nurture a society where technology is a tool, not a crutch, and human relationships are valued and prioritized. This moment in history presents a chance to reevaluate our relationship with technology and to reaffirm the importance of human connection, creativity, and empathy in our learning environments.

"Diverse global, equitable Montessori 2040."

Chapter 74
Circling Back

Reflecting on the impact of large language models, it becomes clear that they encompass far more than the wisdom of renowned thinkers. These models have integrated a broad spectrum of internet content, capturing the diverse contributions of people globally. This reveals that our collective knowledge, creativity, and even the simplest online interactions contribute significantly to the intelligence of these programs. We are not just leveraging the insights of experts; we are benefiting from the shared wisdom of the entire world.

Large language models serve as platforms that amplify all voices, from well-known scholars to anonymous Internet users. This inclusivity ensures that every digital footprint, whether profound or mundane, adds to the collective intelligence that these AI systems represent. While we may not discern the specific impact of everyone's contribution, the cumulative effect is undeniably substantial.

Engaging with these AI programs connects us to humanity's amassed knowledge and experiences, transcending mere technological interaction. This engagement should be approached with gratitude and recognition for the myriad contributions that form the backbone of these systems. They embody not just a single nation's intelligence but the global community's collective intellect.

This new understanding shifts our perspective from idolizing a handful of intellectual giants to recognizing the collective contributions of humanity. Each person's input, whether small or significant, is an integral part of this remarkable advancement in artificial intelligence. The exact measure of everyone's contribution to the vast expanse of internet content might remain unknown, but the collective influence is significant: we are all contributors to this shared digital heritage.

Our every online action is woven into a vast digital tapestry, highlighting the democratization of knowledge in this digital age. This collective consciousness, shaped by our online interactions, continually molds AI. In this context, the ethical implications and the responsibility behind our digital contributions. The quality of what we share online directly influences the development of these intelligent systems. We are not merely beneficiaries of this digital legacy; we are its stewards, responsible for the content we contribute.

Navigating this interconnected digital world presents immense possibilities and challenges, calling for global cooperation and creative contributions. Contributing thoughtfully and ethically, we help shape a legacy within these AI models that future generations can look upon with pride. We stand on the shoulders of a richly diverse and contributing humanity, each playing a role in the ongoing evolution of artificial intelligence. This realization invites us to embrace our role in this digital era, ensuring that our collective journey with AI is marked by wisdom, inclusivity, and responsible stewardship.

"HD photorealistic image of a large group of diverse Montessori children happily weaving a huge colorful tapestry."

Chapter 75
Known

In the realm of education, particularly in Montessori schools, the deep and meaningful connections between teachers and students stand out as a hallmark of the teaching experience. These connections are so profound that students often return years later, seeking references from their elementary teachers for college applications. This unique aspect highlights a critical point in educational philosophy: teachers need to know their students well beyond academic achievements and test scores.

In Montessori environments, this connection is nurtured over a period of three years, allowing teachers to engage with students in a multifaceted way. Teachers observe and interact with their students in various roles and settings during these formative years. This extended time frame offers a comprehensive view of each student's academic, social, and emotional development.

The Montessori method emphasizes individualized learning, where the curriculum is tailored to each student’s needs and interests. Students are encouraged to explore their interests in this setting, leading to a more intrinsic motivation to learn. Teachers observe these explorations, gaining insights into each student's passions, strengths, and areas for growth.

Leadership is a key component of the Montessori curriculum. Students are given numerous opportunities to assume leadership roles, such as mentoring younger classmates, leading group activities, or managing classroom responsibilities. These experiences allow students to develop confidence, responsibility, and a sense of community. Teachers, in turn, witness their students' growth in these roles, understanding their capabilities and personalities more profoundly.

The Montessori approach emphasizes practical life skills and hands-on learning. Students engage in activities that teach them real-world skills, from basic chores to complex problem-solving tasks. Montessori classrooms are a community. Students of different ages work together, fostering a sense of belonging and mutual respect. Students interact with peers. handle conflicts and contribute to the community. Students are seen as whole individuals, each with unique strengths, challenges, and aspirations. Educators are mentors equipped to support their students' academic and personal growth.

Envisioning a future where all educational systems embrace this holistic approach is both ambitious and essential. It would transform schools into vibrant communities where every student is deeply understood and valued, not just for their academic achievements, but for their unique individuality. In such a system, teachers forge meaningful connections with their students, profoundly shaping their educational journeys and equipping them to navigate future challenges with confidence, whether in college, careers, or beyond.

"A powerful, edgy, HD photograph of a diverse and inclusive group of Montessori students roundtable outside in nature"

Chapter 76

Playing Small

One of the things I am interested in is the hierarchy within the education system in England. Positions delineate the hierarchy, from class teachers or assistants to heads of year, departments, or schools, and even the head of finances. These titles are significant, reflecting how we understand and interact with others based on their professional roles.

This leads to a desire to elevate our titles, believing a basic degree is not enough. People pursue higher qualifications, like Master's or Doctorates, for prestige and respect. This chase for titles is driven by an internal conversation about reaching 'there' and is reflected in the school system, where testing and placement in ability groups influence our self-perception and behavior. One might feel demotivated if placed in a lower group, while placement in an advanced group can lead to impostor syndrome or a sense of superiority.

This system creates a divide between winners and losers, which is evident early in schools with gifted and talented programs. Parents often take pride in such labels, seeing them reflect their gene pool. However, these labels can be limiting, defining us in ways that may not reflect our true potential.

With advancements like AI and adaptive technologies, there are now more opportunities for children with learning differences to be seen and heard. Apps and tools are being developed daily to cater to various educational and physical learning differences, promoting educational equality.

Dr. Montessori, a pioneer and visionary, faced many challenges in becoming an engineer and later a doctor. In medicine, she worked in a male-dominated field and had to conduct dissections alone at night due to gender restrictions. Her determination led her to develop the Montessori method, focusing on observing children and tailoring education to their needs.

The Montessori classroom offers a unique and free environment for guides and students, fostering independence and creativity. In the age of AI, Montessori educators are well-placed to integrate these innovative technologies into their environments, using them as a research tool and as an object of research itself.

The education system's focus on labels and titles can be restrictive, influencing how individuals perceive themselves and their capabilities. The Montessori approach offers a unique perspective, encouraging educators and students to transcend these labels and realize their full potential. In doing so, they can avoid 'playing small' and contribute meaningfully to a rapidly evolving educational landscape.

"A child spatial awareness."

Chapter 77

Confusion

Since COVID-19 in 2020, many people have gone online for virtual meetings and conversations, Zoom calls, FaceTime, or WhatsApp, changing how we hold meetings. This change happened suddenly. However, many people have already become accustomed to working virtually for years. It was not new; the number of people engaged in that process was the only difference, plus a few more apps.

In the early days of online calls, if your room were a mess, Zoom gave us background blurring or fake backgrounds so we could hide what was going on behind us and make ourselves look professional. Meanwhile, behind us may be our screaming child, barking dog, or a pile of washing up. This idea of being authentic in confusion has weirdly made us more human. The person leading a meeting now says, "I am dealing with a sick parent, and they may call me, and I might have to go quickly, so please take that into account." Or "My cat loves sitting on the keyboard when they want attention; just to let you know."

We have all found our way into the homes and under the skin of humanity, of human beings in unique and unexpected ways, and in their natural habitats. We often joke that from the waist up, we can have a nice shirt or jacket and look professional, but we could wear pink tracksuit bottoms and big orange fluffy slippers, and it would not matter. During this time, we moved from looking good in polished shoes and fitted suits to being comfortable and authentic. We asked that people include "my mom's visiting, and she might pop in with a cup of tea for me" or "I have a headache, and I might go off screen because I just don't feel well."

We have welcomed humanity into our workplace in a unique way. You cannot guarantee Internet, connection, or sound consistency anywhere. There are many technological challenges in this new way of working. Now, we plan for confusion. We have developed great ways of making people feel comfortable, “You don't have to be on screen if you feel uncomfortable, but we would love to hear your voice.” Now, we have a variety of ways to be included, to show up but not be in the spotlight, and not be excluded.

Will there ever be a time in history when there will not be confusion, where the systems for communicating online will be so completely seamless, so effective that there is never any doubt that we can be seen and heard, and everything will be working? Hopefully, if we get to that point, we will be so used to being completely authentic, including what is going on for us. Yes, we are dealing with a messy house; whatever we are dealing with, it will be a conscious choice, not to look some way on the screen or to try to “look good.”

In a world increasingly entangled in the complexities of deep fakes, the inherent unpredictability and imperfection of human nature may become pivotal in discerning truth from fabrication. It may be through the subtle nuances and idiosyncrasies unique to human behavior and communication that we can begin to navigate and recognize authenticity amidst digital deception.

"A high-definition photorealistic image of a sink overflowing with dirty cups, plates, bowls, pans, and cutlery, in front of a window where plants are dying, you can see the garden outside and washing on the line."

Chapter 78
Summary

There has been a peculiar trend over the last thirty to forty years, more pronounced recently: the rise in soundbite information; news clips, Instagram reels, Facebook stories. We are getting accustomed to digesting information in small bites. We crave summaries, preferring not to read lengthy documents or trawl through academic papers. We want the essence of what is being conveyed, which affords us the freedom to decide whether to spend hours reading a particular newspaper, article, or book.

This trend was highlighted in a presentation by Nita Farahany on AI at the AMS Labposium in San Diego. She spoke about how AI had enabled her, as a professor, inundated with books, articles, and papers, to discern what she wished to spend her time on. With AI summarizing documents, PDFs, and website content, there is now a deeply respectful way of honoring people's time. It is not that we are addicted to soundbites or want everything in a Cliff Notes format, but we now have the choice of how we spend our time. Apps offering digested or short summaries of chapters or entire books are becoming increasingly popular.

This development is exciting because it allows us to feel comfortable with our knowledge base, having had the opportunity to discern where to invest our time, emotional and mental energy, without discounting other perspectives or information. We can now quickly crunch a lot of data and have AI assist us in learning and discovering. This feels like a respectful assistant in the process of gaining knowledge.

If we guide children in this direction, letting them know they have several ways to find information and encouraging them to experiment with these tools, it could transform their learning experience. They might find their mode, whether it is absorbing everything in short bites or delving deeper after listening to summaries. Sharing this with children, respecting them as researchers and lifelong learners, could be incredibly empowering. Imagine if all the books from school were available on an AI app for summarizing and cross-referencing; it would have been a valuable resource for studying history or researching artists.

We are in an era where we have access to all knowledge. As individuals, we need to figure out what works for us, what ignites our passion, and what inspires us to move forward feeling powerful, engaged, and informed. This accessibility to knowledge is a powerful tool for everyone.

"Close up photograph of multiracial children of different faiths Building a robot out of found materials, an HD hyper realistic photo. The sun light is shining as dapples shadows on the children."

Chapter 79

The Adult

When we deal with AI, we look for experts, for the adults in the room, to guide us through the potential challenges and advancements. We seek direction and assurance that the world is being cared for and that things are handled. This reliance on authority figures stems from our upbringing, where parents manage our fundamental needs, making us feel secure. However, there comes a point when we push against this, desiring control, and agency over our lives, be it our living situation, career, or relationships.

We are accustomed to looking for support, wanting those above us to handle things so we can relax. But what if this is an artificial construction? What if society has shaped this reality, where authority figures fend off the horrors of the world, presenting us with a more manageable version? They protect us, allowing us to focus on our work or studies. But can corporations, schools, or institutions genuinely have our best interests at heart?

We trust that our educational system ensures equity and well-being for students. We trust our parents to prioritize our best interests. However, what percentage of that is achieved?

The messiness of being human complicates this trust. We may feel that one parent understands us, not the other, or that their expectations do not align with our aspirations. This leads us to the question: who will be the adult in the room if our families and educational systems fall short? If we are not supported at home, we turn to the next best thing, the education system. But again, how many can say their education fully met their needs and enabled them to thrive?

So, who will guide us in a world increasingly shaped by AI? We tend to trust certain groups, industries, or agencies, believing they provide bulletproof information. But even academics, bound by their research and funding sources, are subject to human nature's subjectivity.

Perhaps, this is not the time to look for external adults or authorities but for us to become that for ourselves. Can we tap into our inner authority and discern what is needed and wanted in each situation? This approach differs from waiting for external salvation, like the Superman illusion, and focuses on being our own superheroes.

Dr. Montessori, grounded, did not endorse cartoons or fantasies. She advocated for a practical approach, encouraging us to be our own guides and authorities. In this age, we can take charge of our narratives and be the adults in our lives, steering our own paths with wisdom and self-assurance.

“Diverse photorealistic young girl works with blockchain in a bright open classroom.”

Chapter 80
Handles

When did "handles" go from being objects to pull yourself up by or hold onto for stability to a name you call yourself on an intercom or phone? It is interesting how words get re-adopted or created from nowhere, like "Google" or "Tweet." As human beings, we are hard-wired to adapt quickly to new uses of words and ways of doing things. Our adaptability is extraordinary, something Dr. Montessori researched extensively.

We are also quick to normalize new situations, like our sense of smell alerting us to something burning or the aroma of food, then becoming accustomed to it once the need for the alert passes. In AI and education, we normalize things that seem off.

Our biological system notices things to alert us to danger, but we adapt quickly and make it normal. The same is true for AI. Initially, we were alerted to the power and influence of companies like OpenAI, questioning their safety and ethics. However, as these technologies become more commonplace and others do not seem concerned, our internal alert systems can switch off. We might ignore our concerns and agree with the majority, which is not always the best action. Ignoring intuitive knowledge about physical or mental well-being can lead to larger issues.

As humans, we have created a narrative where authority figures have everything under control, leading us to question our concerns. We have normalized by ignoring our internal alert mechanisms. Yet, there are people, the "canaries in the coal mine," raising alerts about honoring children in education and the risks of losing a generation to technology. These voices are crucial in a world increasingly intertwined with technology.

In the context of technology and AI, handles are there for support to help us achieve more than we could on our own. If we hold onto the handle of AI and technology, using it to strengthen ourselves, that is where the difference will be made. We do not need to be passive consumers of technology but active participants, using it to enhance our capabilities and understanding.

The evolution of language around technology, like “handles,” reflects our changing relationship with it and the need to adapt while staying mindful of our internal guidance.

"Diverse global universal multicultural creative expressive Montessori 2060."

Chapter 81

How Long

How long will it take for the world to change now that we have AI? How long until businesses make it mandatory for their staff to use AI, or when will it be built into every system, with every individual having their personal AI on their computer to learn about them? When will this become a common experience? How long until you need to be proficient in AI, even trained in AI, to secure a job? Will it become necessary to have AI credentials to work in a company that has declared 100% AI use?

How long will it take for schools to integrate AI into regular computer programming systems for students? When will it become automatic for students to have access to programs that accommodate their learning and physical differences? I wonder how long this change will take. How will the world look once it becomes assumed that you will be working with AI in companies and schools? Will it free people to have half-day work schedules because everything is more efficient? Could schools finish at lunchtime, allowing students to pursue other activities in the afternoon?

Is it a matter of time changing because of the efficiencies AI brings? Are companies using AI now functioning more optimally, needing less staff but making more money due to effective advertising? Is AI automatically seen as the key to success? If so, how much will companies rely on their employees to be original, creative, and experimental? With routine jobs handled efficiently by AI, more energy could be channeled into innovation, creativity, and experimentation. Those who can use AI in these more imaginative ways will be better compensated.

We now talk about the creator economy, where everyone has the tools to be incredibly creative. This shift mirrors the Industrial Revolution, which automated many physical and manual jobs. This led to concerns about craftsmanship, prompting William Morris in England to create the Society of Designer Craftsmen to preserve beautiful, handcrafted objects and human input.

So, if AI takes over day-to-day jobs or speeds them up, it could free us to be craftsmen, creators, and innovators. The world might become more diverse, beautiful, unique, and unusual as people bring their visions and ideas to life. We could look forward to a colorful, creative, and beautiful world. But, of course, this is contingent on how well everything ends with integrating AI into our daily lives.

"Larry Paige, Sergei Brin, Jeff Bezos, and Bill Gates, as kids, in a Montessori early childhood classroom."

Chapter 82

Flooded

The fascinating aspect of AI programs lies in their prolific nature. When you pose a simple question to an AI, it often responds with more information than anticipated, sparking further curiosity and exploration. This can lead to the accumulation of documents. Thankfully, AI programs typically feature a history panel, allowing you to trace your thought process and manage the information. However, for those who are prolific thinkers and creators, the accumulation of documents and ideas can become overwhelming.

Consider Midjourney as an example. This AI tool creates four images based on your prompt. You can then expand and modify these images, leading to an extensive collection of visuals. For those engaged in creative projects excitement can lead to a vast hoard of data. The challenge lies in knowing when to stop, especially when creativity is at its peak.

The question then arises: where to store all these AI-generated documents and images? Is it more practical to save them on your computer, or should they remain on the AI platform? While these platforms are designed to manage enormous amounts of storage, personal collections can quickly grow into hundreds or thousands.

When you extrapolate this to a broader scale, considering the exponential growth of AI usage across various sectors, the situation becomes even more pronounced. In fields like advertising, where AI is employed for a multitude of ideas, images, and iterations, the information flood is immense. We find ourselves in an era where access to original ideas and unique insights is unparalleled, but this also requires a strategic approach to their utilization.

The critical challenge in this age of information overload is effective management. It is not merely about discerning which documents are valuable, but also knowing how to utilize them effectively. Our approach to managing this influx is essential. It is no longer just about the ability to create AI images or documents; the true skill lies in managing this information in a way that enhances your personal and professional life.

Moreover, there is the aspect of trust in AI systems. Can we rely on these platforms to persist over the long term? This consideration is crucial, particularly when contemplating revisiting an idea or image in the future. It raises questions about personal data management strategies, which go beyond mere technical solutions to encompass unique, individual approaches to managing vast amounts of information.

In navigating the flood, maintaining integrity, and developing effective data management skills are paramount. We must be able to trust the systems we use, ensuring the longevity and accessibility of our creative outputs. This situation calls for a balance between embracing the opportunities AI presents and maintaining control over the deluge of data, ensuring that our interactions with AI remain productive, manageable, and enriching.

"An HD photo of water, full of digital information flowing powerfully at full force towards a young diverse teacher."

Chapter 83

Under the Bus

Social apps like YouTube have enabled those who wish to voice their opinions in countries where such expressions are frowned upon, even risking personal harm, to do so. This highlights a situation where traditional research and investigation methods are paralleled by younger, innovative thinkers eager to conduct their research, gather data, stake their names on their findings, and make a point as citizen reporters and scientists.

Emerging newsletters like TLDR, Ben Bites, and The Sequence offer digested updates on AI advancements, including regulatory changes, actions by major companies, and new innovations. They condense information into digestible segments, helping us stay informed in a rapidly evolving landscape. This information, however, is as fleeting as the present moment; new developments can quickly redirect the conversation. We live in a time of rapid, unpredictable changes, which some may perceive as irresponsible or unreliable.

We have been conditioned to align our understanding within a structured framework, seeking order and clarity as traditionally taught in educational settings. Yet, the influx of new forms of information, carrying equal weight and respect, is reshaping this landscape. Information once confined to private emails or documents shared among industry insiders is now becoming more publicly accessible, offering real-time insights into the dynamics of businesses and AI development.

The freedom of information era is revealing, yet we are only scratching the surface. Beneath the visible lies a vast reservoir of inaccessible knowledge. Historically, the risk of speaking out, especially in academic circles where career and funding were at stake, was tangible. Individuals risked professional ostracism for expressing unconventional views, and to be thrown "Under the Buse" for convenience's sake when information was unwelcome.

As we adapt to this era of information abundance, discernment becomes crucial as we learn to sift through the vast amounts of data, identifying what is relevant and valuable. The skill of discernment is not just about filtering information but also about understanding its context and potential impact. It is about making informed decisions in a landscape where information is abundant but not always accurate or beneficial, but how to we teach this vital skill to children?

The responsibility of sharing information has shifted. No longer is it solely the domain of academics or industry experts; it has become a collective endeavor. As our children learn to share and consume information, how do they gain a sense of responsibility, understanding the weight their words and insights carry?

As we navigate the "Under the Bus" era, our focus should not only be on the quantity of information but also its quality and our interaction with it. By our children cultivating discernment, responsibility, and trust, they can harness the power of information for positive change, creating a world that is well-informed, coherent, and effective.

"Illustrate the metaphor of " Being Thrown Under the Bus" as it relates to schools and information. Make an HD and photographic image."

Chapter 84
Cartooned

Spending time on AI apps like Midjourney reveals a striking cartoon nature to the images, especially if high-definition photorealism is not specified. This trend reflects the social media vortex we are engulfed in, where there is a compelling desire to portray ourselves as cute, beautiful, sexy, and ensconced in opulent backgrounds. This superficiality is vividly apparent in Midjourney, where users generate prompts, leading to a vast array of iterations from a multitude of users. The output often gravitates towards aesthetically pleasing images, from cute women, and handsome men, fulfilling specific visual desires or personal preferences of the creators.

This gradual shift towards a cartooned world is not limited to personal use but extends into professional environments. Companies often succumb to the allure of adding engaging, cute emojis or delightful cartoon images in PowerPoint presentations, posters, and advertising campaigns. This strategy is employed to lighten the mood, entertain, or cleverly manipulate the audience's perception. Cartoons have become a well-established medium for inducing emotion, even in serious, information-heavy presentations.

Children have been deeply immersed in the world of cartoons for many years. Television shows, magazines, and various forms of children's entertainment are replete with animated characters, from Dora the Explorer to myriad other cartoon figures that have become staples in young lives. This constant exposure has ingrained a preference for cartoon-style representation from an early age.

This contrasts sharply with the Montessori method, which is rooted firmly in reality. Montessori educational materials, such as three-part cards, use actual images, such as photographs of food items, to teach food groups. These are often paired with real foods or accurate plastic replicas for tactile engagement.

Dr. Montessori's approach avoids using cartoons in teaching materials, newsletters, or websites, adhering to her principle of engaging children with real-world experiences. This philosophy extends to avoiding dressing-up corners with princesses and heroes; even play kitchens are designed to follow reality as closely as possible.

As our world increasingly embraces the cartooned aesthetic, with even real movies being transformed into cartoons through apps, we face a crucial decision. Do we want to live in a cartoon-dominated world or seek to engage with the natural world?

Choosing authentic experiences is more relevant than ever. While cartoons and digital animations offer a form of escapism and creative expression, the Montessori method highlights the significance of real-world interactions and learning. It suggests that engaging with the world can foster a deeper understanding and appreciation of our surroundings, leading to more meaningful experiences and learning outcomes.

"An HD photo of a young female swimmer with short brown hair. Swimming in the pool with oversized colorful goggles, smiling and happy."

Chapter 85
Action

One of the key aspects of the Montessori method is that students are naturally inclined to take action, thanks to the freedom of movement and thought they enjoy. This freedom allows them to follow intrinsic motivations and internal guidance. From the earliest stages, even infants in Montessori settings are given autonomy. They sleep in floor beds, allowing them to crawl in and out as needed, fostering a deep sense of self-reliance and awareness of their needs.

In contrast, traditional child-rearing practices often involve managing children's movements, from placing them in high cribs to moving them around as necessary. This approach can inadvertently undermine a child's sense of autonomy and instinct for self-direction. In Montessori environments, from an early age, children are encouraged to exercise their autonomy, choosing activities that interest them and following through with their chosen tasks.

In Montessori, this foundation of autonomy and intrinsic motivation is well-established when children reach early childhood. Children develop reading and math skills in these classrooms, building on the foundations in the infant-toddler room. They exhibit a significant degree of free will, clearly expressing their interests and preferences, which are deeply respected by educators.

However, we often create barriers to taking action in the wider world. Whether it is due to fear of failure or societal norms that dictate what is or is not appropriate, many of us have inhibited our natural instincts to act. This inhibition is contrasted with the development encouraged in Montessori settings, where autonomy and action are nurtured from an early age.

The question then arises: what do we need to implement or remove in our education systems to foster this level of agency and autonomy? Why does it matter? We live in a world where action drives results. The advancements we see in AI and technology result from people taking action on their ideas and beliefs. If our education system suppresses this instinct to act, we risk stifling innovation and creativity.

The Montessori method offers a blueprint for an education system where natural instincts for well-considered, thoughtful, humanistic, and balanced actions are cultivated. Children learn to see the common good and work within that framework, taking actions that build upon and improve existing structures or create new ones with integrity.

There is a compelling case for reevaluating our education systems to promote the instinct for action. By doing so, we can prepare future generations to contribute positively and innovatively to society, embodying the true spirit of action that drives progress and change.

"Light, neutral, simple, unexpected, Asian, diverse, inclusive, nature, Montessori, background, soft focus, inspirational, cultural."

Chapter 86

Savings

The concept of savings was instilled in me from an early age. I remember having a money box and gradually transitioning to a checkbook and a savings account. However, in the gig economy, the notion of consistent pay or salary, and thus the idea of savings, seems almost far-fetched. It starkly contrasts with earlier times, prompting me to think about how we view education and our investment in human capital.

Our children represent a sizable portion of our nation's and the world's wealth, primarily through education. The funds we channel into public schools are a savings account. We invest in buildings, curricula, teachers, and more importantly, our children. The hope is that this investment pays off, which it often does. Innovations like AI are testaments to the success of our educational investments.

However, it might be less optimistic. Global systems are struggling, as evidenced during and after COVID-19, with logistical programs crashing and failing to recover adequately. Coupled with a culture of secrecy and a reluctance to admit faults, we lack a clear picture of how well our systems and economy function.

When considering our children as an investment, we must question how we perceive the return on this investment. How do the actions of millennials and subsequent generations reflect the value of our educational contributions? While traditional values like securing a decent job, getting married, and owning a home still hold, there is an emerging trend of valuing relationships and experiences more, often leading to a nomadic lifestyle.

This shift in values prompts a reevaluation of our investments in education. Are we truly getting the benefits we hoped for? As some Montessori training programs move online, offering convenience and flexibility, we must question if this shift affects the quality of education, and the caliber of teachers produced.

Investment in human resources, starting with children, is akin to saving for the future. But if the outcome of our educational investments is not producing individuals who are autonomous, thoughtful, and act with integrity and intention for the common good, we need to reassess our strategies. It is not enough to save pennies; we need substantial investments to yield significant benefits.

So, if our children are our savings scheme for the future, are we doing a good job? It is time to review our educational processes and change our approach. Like altering a piggy bank, we must adapt our investment strategies in education to ensure they align with our society's evolving values and needs. Only then can we expect a meaningful return on our investment, reflected in the caliber and character of the next generation.

"A large white piggy bank on a table, with real multicultural, diverse children being dropped into the coin slot."

Chapter 87

Safe

Public schools in America invest significantly in safety protocols, ensuring necessities like fire systems, well-structured buildings, and safe electrical and water systems. Beyond these, schools also focus on ensuring students, teachers, and staff are aware of safety protocols. This comprehensive approach extends to the public system, reflecting our deep concern for safety.

However, since the advent of school shootings, additional measures like police presence and scanners at school entrances have been implemented, particularly in middle and high schools. These measures aim to prevent children from bringing in weapons, further strengthening the safety net around students. Yet, paradoxically, the more we bring external safety measures, the more trapped and imprisoned students might feel, protected from life and each other.

This over-protection does not always address the root causes of why children might feel unsafe, such as mental health issues, bullying, or academic pressure. Often, those who commit acts of violence in schools have themselves felt unsafe and then project that feeling onto others. In our quest to protect, are we neglecting to instruct children about natural consequences?

In many school settings, children are shielded from potential hazards, whether it is using scissors, being near an open flame, or handling glass containers. In contrast, Montessori schools advocate for learning through natural experiences, even if it involves some level of risk. Children learn best through their own experiences in an environment where they understand and navigate risks safely.

With the advent of artificial intelligence, the challenge of ensuring safety becomes even more complex. Schools might unwittingly implement AI programs that negatively impact children, or a child might inadvertently expose themselves to danger while exploring online.

In this new world, instead of incessantly trying to externally protect children, can we teach them to protect themselves, to be agents for their own safety? If we wait passively until they are 'old enough' to handle safety, we miss out on developing their capacity to make informed, capable decisions.

If we underestimate children's capabilities, we lose out on making them partners in safety. This includes supporting children who struggle with feelings of anger and grievance, encompassing issues like bullying, academic pressure, and inclusivity for those with learning differences. Often, children with learning differences have been pulled out of regular classes, making them feel singled out and emotionally unsafe. True safety might lie in creating environments of true equity, where children feel they can function happily and productively in any setting. This approach might lead to the safest environment we have ever seen, one where children are empowered, aware, and actively involved in maintaining their own safety.

"A children's open sketchbook with sticker figure sketches of school. Colored and pages filled."

Chapter 88

Theory of Multiple Intelligences

In the tapestry of human cognition, Howard Gardner's theory of multiple intelligences offers a vibrant and diverse palette, each color representing a unique aspect of our intellectual capabilities. As we advance into an age increasingly shaped by artificial intelligence, understanding these varied intelligences becomes not just a matter of academic interest, but a crucial tool for personal and societal growth.

The Spectrum of Human Potential

Bodily-Kinesthetic Intelligence: Do you express ideas or emotions through physical movement? This intelligence reflects the ability to control bodily motions and skillfully handle objects. It is about using the body in a highly differentiated and skilled way.

Interpersonal Intelligence: Are you particularly good at understanding and interacting with others? This intelligence revolves around the capacity to understand the intentions, motivations, and desires of other people and work effectively with them.

Intrapersonal Intelligence: Are you introspective and self-aware? This form of intelligence pertains to the understanding of oneself, including one's own feelings, fears, and motivations.

Linguistic Intelligence: Do you have a way with words? People with linguistic intelligence find it easy to express themselves verbally and are typically good at reading, writing, storytelling, and memorizing words.

Logical-Mathematical Intelligence: Do you have a knack for reasoning, recognizing patterns, and logically analyzing problems? This intelligence involves calculation, logic, and the systematic exploration of abstract patterns.

Musical Intelligence: Are rhythms and melodies like a second language to you? This intelligence embodies the ability to recognize and create musical pitches, tones, and rhythms.

Naturalistic Intelligence: Do you feel a deep affinity with the natural world? This intelligence is about understanding living things and reading the natural environment, recognizing flora and fauna, and making use of such knowledge.

Visual-Spatial Intelligence: Do you think in images and pictures, and are you able to visualize accurately and abstractly? This intelligence is about spatial judgment and the ability to visualize with the mind's eye.

As artificial intelligence reshapes the boundaries of what machines can do, the uniqueness of our human intelligences becomes increasingly significant. These intelligences are not just about skills or abilities; they shape how we view the world, interact with others, and find our place in the AI-driven era. These intelligences underscore our unique human qualities, the ones that make us irreplaceable, creative, and deeply connected to the world and each other.

"A close-up, HD hyper realistic photographic impression of a young child exploring the Forest floor, the same size as the roots, mosses, mushrooms, and other fauna and flora. It is late afternoon, and the sun is glinting through the tree canopy. The light glistens on the wet of the forest floor."

Chapter 89
Exciting

As you have seen from this series of musings, I find myself somewhere between a six-week-old puppy and apocalyptic regarding artificial intelligence. This is where I feel confident to plant my flag. I have been reading and researching, keeping my eyes open and my ears to the ground. I am fascinated by how humans have developed this incredible technology that makes me feel exponentially more capable, creative, powerful, and motivated to use all my talents and gifts.

Using the recording process instead of writing has been a game-changer for me. As much as I love two-finger typing, my fingers cannot keep up with my brain, which works incredibly fast. Not fast in a boastful sense, but fast in that ideas come thick and fast, and thoughts change direction rapidly. I have so much to express, wonder about, and investigate, and my brain works in such a way; at school, I was never any problem for my teachers because by then, I had learned that silence was my best option.

Until ChatGPT 4, Claude 3, and Midjourney, I had not found a way to express all these ideas and explore my varied interests. At art college, my work focused on the place of women in society and how we can become invisible. I worked with clay, metal, and wood. Now, I see AI as my next material. I do not need to replace other materials; it is just another tool. I can keep moving it through iterations digitally, creating 3D digital sculptures and images.

What is so funny is the copious number of iterations possible in a concise space, which helps move my idea exponentially forward. Just a little bit of a nuance change, a word change, or an adjective, suddenly it is completely different. I have also noticed that it starts speaking in my voice through ChatGPT 4, Claude 3, or Midjourney. I see myself expressed in the work, my learnings, my interests. Whether it is spooky or goes back to some AI mythic way of being, it feels like it is responding to me personally and listening to me, which we will not delve too deeply into here; that is a conversation for another book.

So, we have a lot to discuss and unpack during AI's emergence into the world. I wish you all fabulous, deep conversations and exciting moments. Whatever your fear and concern, I earnestly request that you take that into the world. Keep asking your questions and seeking answers. I offer you the time to be true to yourself, to your intelligences, whatever they may be, and that you find answers for yourself in your life that bring you creativity, joy, contentment, and advancement. Create a life that thrills you, one that you love, and honestly reveals and matches your Montessori "cosmic purpose," seen, respected, and loved for your visions, ideas, and heart.

"A diverse photorealistic happy child plays with blockchain in a winter wonderland."

Chapter 90

Non-Linear Ending

Thank you for journeying to the end of this book. Thoughts, as we all experience, are non-linear. They dart and dance into our consciousness randomly throughout the day, big and small, good, and bad, scary, and happy. They pop up unexpectedly, demanding our attention to varying degrees. Thoughts may hold us captive, pulling at our minds like a persistent loose thread. We may find ourselves returning to them, unraveling them further into a larger tapestry of contemplation.

I leave these thoughts with you, allowing them to flow and be as fleeting or as persistent as they naturally are. I invite you to weave your insights, opinions, and assessments into this ongoing dialogue. Your perspectives enrich the collective understanding and deepen the conversation, particularly in the realms of AI and education.

Silence, often seen as the absence of communication, can indeed convey volumes. It can signify deep resignation, a quiet refusal to engage, or an inability to find the right words. The silence surrounding these topics might speak of the overwhelming nature of our technological future and the vastness of our educational challenges. Your journey through these pages is a testament to your willingness to engage with these complex issues.

In closing, I extend my deepest gratitude for your time, attention, and the generosity of spirit you have brought to this journey. As we continue to navigate the ever-changing landscape of technology and its impact on our lives and learning, your participation and engagement are invaluable.

Together, let us strive to bring balance to a world increasingly shaped by these powerful and transformative forces.

With my deepest love and respect,

Kate

"A HD hyper realistic, close-up image of a happy, heavily pierced and tattooed, biracial, cyborg grandmother, with closed eyes, being gathered leaf by leaf up suspended into the air and entwined by fern leaves, in the midst of beautiful, warm, green, woodland, forest scene. There should be close-up leaves and branches and light beams from the sun piercing through the leaves of the plants and water droplets in the air. White, blue, and green smoke spirals around the scene lending an air of mystery and wonder."

Chapter 91

Acknowledgments

In the creation of this book, a journey both exhilarating and challenging, I have been fortunate to have a galaxy of stars guiding and supporting me. It is with immense gratitude that I extend my thanks to each one of them, if I have not thanked you by name, please include yourself in this acknowledgment with my love.

First and foremost, I am eternally indebted to Dr. Betsy Coe, who sadly passed away in 2023. Her steadfast commitment to the Montessori method was a deep inspiration to me as I was undertaking my Elementary I, II, and Administrator training at the Houston Montessori Center (2008-2016). To Cofounders of the TIES M. Ed program Dr. Philip Snow Gang and Marsha Snow Gang (also passed), for the freedom to experiment, which gave me the academic confidence I needed to use all my intelligences.

To all the brilliant leaders and guides within UK public schools that I worked in and US Montessori schools, and most importantly to every individual who participated in the creation and gentle nurturing of Kingwood Montessori School. Your generosity, professionalism, wisdom, and insight have touched my heart and deeply inspired my journey and these musings.

Immeasurable love and gratitude go to my precious family. To my sister Susan for being by my side throughout our nomadic childhood, to my father Brian and mother Anne, (both passed) for your brilliance, creativity, and wisdom.

To my dearest childhood friend and most powerful inspiration Astrid Longhurst, for playing with me so gracefully over these many years, and for always seeing the invisible with me.

For my brilliant, dedicated, fun, handsome, and loving craftsman of a husband Mark, who has stood firmly beside me in many exciting iterations of our lives, enduring my long hours of seclusion with support, understanding and endless cups of tea, providing me with the peace and space needed to write, I love you deeply. To our extraordinarily intentional, insightful, unique, and inspirational children Amelia, Olivia, and Alexander. For you, taking the delicate seeds of your unusual Montessori childhood, tending to them gently, and enabling them to flourish and grow into a world which is authentic to each one of you. You are our pride and joy.

I extend my sincere thanks to the American Montessori Society leaders, Gina Lofquist, Dean Blasé, Rosemary Quaranta, Gabrielle Holt, and Hannah Baynham, for supporting the research of the first and second cohorts of the Innovation Fellowship Think Tank in 2023-2025, and love and thanks go to Eric Rodriquez, formerly Intel now global keynote speaker on being human in the age of AI, for setting the pace and always moving the bar higher.

I am profoundly grateful to my editor, Anna Shildrick, education advisor, and co-founder of AI Education Connective, whose keen eye and expert guidance helped refine my words and ideas. Her suggestions were invaluable in bringing clarity and precision to my work.

"An HD hyper realistic, close-up image of an Ai-enhanced, cyborg teenager, showing their head and hands as they happily, and excitedly explore in the midst of beautiful and colorful tropical flowers plants, birds, snakes, animals, and insects. of There should be light beams from the sun piercing through the leaves of the plants and pollen and water droplets floating in the air."

Chapter 92

Programs & Applications

Alice App AI (iPhone) – Transcription AI

Atticus – Book Editing App

Midjourney AI – Image generator from prompt

Open AI - ChatGPT 3.1, 4.1 & 5.1 - Edited Voice Recorder transcriptions for punctuation and spelling

Voice Recorder App (iPhone) - Record and transcribe

"A hyper realistic, photographic close-up of a beautiful, lively, intelligent old lady in a colorful hippy sweater, illustrating a children's book, with an old-fashioned pen and ink, and bright pots of paint and brushes on the table. She looks happy and thoughtful. She is next to an open window where lots of sweet, colorful English birds are sitting on the windowsill, and all around her, the sun is shining through the window, and dust and water droplets are caught in the light. The image is ultra detailed, the wind is blowing a slight breeze through the windows and the old lady's grey curly hair is blowing freely."

Chapter 93

About the Author

Kate Broughton, M.Ed., is an educator, artist, and practitioner-researcher with more than forty years of experience across diverse educational settings in the United Kingdom and the United States. Her work spans public education, private Montessori institutions, teacher education, nonprofit school leadership, and educator-led research, with a sustained focus on child development, embodied learning, and innovation in education.

Kate holds American Montessori Society (AMS) Elementary I & II and Administrator credentials, as well as a master's degree in Montessori Integrative Learning from TIES in partnership with Endicott College. In 2007, she co-founded Kingwood Montessori School, a nonprofit Montessori institution in Texas, where she later served as Director and as a member of the Board of Directors. Her leadership contributed to the development of a learning environment grounded in Montessori philosophy and informed by contemporary research.

Throughout her career, Kate has contributed to Montessori teacher education as faculty at the Gulf Coast Montessori Teacher Education Center and the Montessori Education Center of the Rockies. She has also served as an examiner for Pearson Education in the United Kingdom, supporting curriculum development and assessment within international education contexts.

In 2023, Kate joined the inaugural cohort of the AMS Innovation Fellowship Think Tank, completing the two-year fellowship, and awarded an AMS Impact Medal, in 2025. Her research, *Mapping Montessori Materials for AI Competency Development*, recognized as a finalist in the 2025 BookFest Awards, offers a developmentally grounded framework for understanding how Montessori environments cultivate competencies relevant to emerging technologies while preserving human agency, judgment, and creativity.

Beyond Montessori, Kate has served in advisory roles related to leadership and the responsible integration of artificial intelligence, including work with the Strategic AI programs at Mercy University and the University of San Francisco. Her advisory work emphasizes human dignity, ethical clarity, and long-term societal well-being in conversations about innovation and the future of work.

Kate's background in the arts continues to inform her of an interdisciplinary approach to education. She holds a BA (Hons) degree in Fine Arts, was a Licentiateship with the Society of Designer Craftsmen (1981), and maintains an active creative practice that bridges traditional craftsmanship with emerging technologies.

She is the author of *Montessori & AI* (Volumes I and II), *Mapping Montessori Materials for AI Competency Development*, *After Alpha*, and *Create Montessori*. Her writing engages educators, leaders, and families seeking clarity, coherence, and confidence in a time of rapid technological change.

Kate lives and works in Colorado.

"A hyper photorealistic image of Montessori school teenagers studying a virtual image of a human heart. They are reviewing the circulatory system and looking closely at holographic images. One young lady is on a laptop looking at holographic images of the lungs, the images floats in front of them and they are using their fingers to direct the image."

Chapter 94

Dr. Maria Montessori

The last word:

> An education capable of saving humanity is no small undertaking; it involves the spiritual development of man, the enhancement of his value as an individual, and the preparation of young people to understand the times in which they live.
>
> **Maria Montessori** | Education and Peace (AMI)

"An HD, hyper-realistic, clean scene, deep in the Amazon jungle, there is the mother tree; she is richly covered in dark green leaves and vines. Light beams through the canopy above. Behind the tree is a stunning luminous blue waterfall that dazzles in the sunlight."

Made in the USA
Las Vegas, NV
03 November 2026

57423764R00228